Management for Professionals

For further volumes
http://www.springer.com/series/10101

K. Ganesh • Sanjay Mohapatra
S. P. Anbuudayasankar • P. Sivakumar

Enterprise Resource Planning

Fundamentals of Design and Implementation

K. Ganesh
Supply Chain Management - Center of Competence
McKinsey Knowledge Center (McKC) |
McKinsey & Company, Inc.
Ascendas International Tech Park
TamilNadu
India

Sanjay Mohapatra
Xavier Institute of Management
Bhubaneswar
India

S. P. Anbuudayasankar
Department of Mechanical Engineering
Amrita School of Engineering
Coimbatore
India

P. Sivakumar
Vickram College of Engineering
Enathi, Sivagangai
India

ISSN 2192-8096   ISSN 2192-810X (electronic)
ISBN 978-3-319-38211-1   ISBN 978-3-319-05927-3 (eBook)
DOI 10.1007/978-3-319-05927-3
Springer Cham Heidelberg New York Dordrecht London

© Springer International Publishing Switzerland 2014
Softcover reprint of the hardcover 1st edition 2014
This work is subject to copyright. All rights are reserved by the Publisher, whether the whole or part of the material is concerned, specifically the rights of translation, reprinting, reuse of illustrations, recitation, broadcasting, reproduction on microfilms or in any other physical way, and transmission or information storage and retrieval, electronic adaptation, computer software, or by similar or dissimilar methodology now known or hereafter developed. Exempted from this legal reservation are brief excerpts in connection with reviews or scholarly analysis or material supplied specifically for the purpose of being entered and executed on a computer system, for exclusive use by the purchaser of the work. Duplication of this publication or parts thereof is permitted only under the provisions of the Copyright Law of the Publisher's location, in its current version, and permission for use must always be obtained from Springer. Permissions for use may be obtained through RightsLink at the Copyright Clearance Center. Violations are liable to prosecution under the respective Copyright Law.
The use of general descriptive names, registered names, trademarks, service marks, etc. in this publication does not imply, even in the absence of a specific statement, that such names are exempt from the relevant protective laws and regulations and therefore free for general use.
While the advice and information in this book are believed to be true and accurate at the date of publication, neither the authors nor the editors nor the publisher can accept any legal responsibility for any errors or omissions that may be made. The publisher makes no warranty, express or implied, with respect to the material contained herein.

Printed on acid-free paper

Springer is part of Springer Science+Business Media (www.springer.com)

Preface

This book is designed to meet the needs of researchers and students. The text assumes that the reader knows basic system analysis and design (SAD). The SAD techniques required for understanding some advanced topics are incorporated in individual chapters.

This book's main objective is to introduce, in a unique manner, the fundamental principles of understanding business requirements and fitting enterprise resource planning (ERP) to meet these business needs. The book also helps to understand the usage of ERP for monitoring and controlling business processes. A sufficient number of topics has been covered in detail to enable the readers to follow without much difficulty.

The first three chapters of this book introduce multiple deployment considerations, project initiation and core process analysis. These chapters provide an overview of prerequisites, activities, deliverables and critical success factors of project initiations and core process analysis. We believe that once you understand the basic principles of detailed activities, creating solutions to critical success factors is a matter of fact.

Chapters 4 through 6 present techniques and strategies for conference room pilot, customization, interface and conversion for improving aligning ERP with business models.

Chapters 7 through 11 suggest system integration testing, training and user acceptance test. These chapters offer recommendations regarding effective and productive testing techniques before production goes live.

Finally, Chap. 12 explores project management and its objectives. This last chapter provides detailed activities and deliverables for project management. We also discuss decision matrices and critical success factors for implementation.

Contents

1   **ERP as a Business Enabler** 1
    1.1   Introduction 1
    1.2   Evolution of ERP 2
            1.2.1   Scope for MRP-I 3
            1.2.2   MRP-II 4
    1.3   Need for ERP 6
    1.4   Overview of ERP 7
    1.5   Modules of ERP 8
            1.5.1   Components of Material Management Module 8
    1.6   ERP Life Cycle 9
    1.7   How ERP Improves Productivity of Business Processes 11
    1.8   Benefits of Automation Through ERP 12
            1.8.1   Collaboration Approach 12
            1.8.2   Cross Functional Barrier is Broken 13
            1.8.3   Integration of Island of Automation 13
            1.8.4   Key Principles for Integration of Islands 14
    1.9   ERP Products 17
            1.9.1   SAP 17
            1.9.2   PeopleSoft 18
            1.9.3   Oracle 18
            1.9.4   Microsoft Dynamics 19
            1.9.5   Comparison of ERP Packages 19
    1.10 Trend in ERP: Open-Source ERP 25
            1.10.1 Definition of Open Source 26
            1.10.2 Advantages of Using Open-Source ERP 26
            1.10.3 Conclusion 27
    1.11 MIS and ERP 29
    1.12 Summary 30
    1.13 Glossary 30
    1.14 Review Questions 31
    1.15 Project Work 31
    1.16 Case Study 32

		1.16.1 Background	32
		1.16.2 Current Situation	33
		1.16.3 Challenges	34
		1.16.4 Discussion Points	34
		1.16.5 Notes that will help in Discussion	34
	Further Reading	35

2	**Introduction**		37
	2.1	Target Audience	39
		2.1.1 Multiple Deployment Sites/Multiphase Considerations	39
		2.1.2 Structure of the Document	40
	2.2	Project Initiation	41
		2.2.1 Objective	41
		2.2.2 Prerequisites	43
		2.2.3 Detail Activities	44
		2.2.4 Decision Matrix/Checklist	46
		2.2.5 Critical Success Factors	47

3	**Core Process Analysis**		49
	3.1	Objective	49
	3.2	Prerequisites	50
		3.2.1 Business Requirement Definition	50
		3.2.2 Business Requirement Mapping	50
	3.3	Detail Activities	51
		3.3.1 Business Requirement Definition	51
		3.3.2 Core Process Analysis: Key Process Owners Training	56
		3.3.3 Core Process Analysis: Business Requirement Mapping and Gap Analysis/Resolution	59
	3.4	Deliverables (Table 3.11)	67
	3.5	Decision Matrix/Checklist (Table 3.12)	67
	3.6	Critical Success Factors	67

4	**Conference Room Pilot**		71
	4.1	Objective	71
	4.2	Business Flow Diagram: Conference Room Pilot	73
	4.3	Prerequisite	73
	4.4	Detailed Activities	74
		4.4.1 Prepare Instance Strategy	74
		4.4.2 Prepare Backup Strategy	74
		4.4.3 Define Application Instance Management Procedures	74
	4.5	Install the CRP Instance	75
		4.5.1 Define Application Setup	75
		4.5.2 Unit Testing and CRP	78
	4.6	Deliverables (Table 4.2)	82

Contents ix

4.7	Decision Matrix/Checklist (Table 4.3)	82
4.8	Critical Success Factors	82

5 Customizations 85
- 5.1 Objective 85
- 5.2 Process Flow Diagram: Implementation of Customization 85
- 5.3 Implementation Schedule—Customization 85
- 5.4 Prerequisite 86
- 5.5 Detail Activities 87
  - 5.5.1 Gap Analysis 87
  - 5.5.2 Analyze Dependency 87
  - 5.5.3 Design 88
  - 5.5.4 Development 88
- 5.6 Deliverables 89
- 5.7 Decision Matrix/Checklist 89
- 5.8 Critical Success Factors 89

6 Interface and Conversion 91
- 6.1 Objective 91
- 6.2 Business Flow Diagram 91
- 6.3 Implementation Schedule—Interface 93
- 6.4 Prerequisite 93
- 6.5 Detail Activities 93
  - 6.5.1 Identify Interfaces and Tools 93
  - 6.5.2 Prepare Interface-Building Plan 94
  - 6.5.3 Interface Designing 94
  - 6.5.4 Interface Development 95
  - 6.5.5 Interface Integration Testing 96
  - 6.5.6 User Acceptance Testing 96
  - 6.5.7 Identify and Freeze Source Data Files 97
  - 6.5.8 Load and Validate Source File for Data Migration 97
  - 6.5.9 Load and Validate Backlog Transaction Data 98
- 6.6 Deliverables 98
- 6.7 Decision Matrix/Checklist 99
- 6.8 Critical Success Factors 99

7 System Integration Testing 101
- 7.1 Objective 101
- 7.2 Business Flow Diagram 101
- 7.3 Implementation Schedule—System and Regression Testing (Table 7.1) 102
- 7.4 Prerequisite 102
- 7.5 Detail Activities 103
  - 7.5.1 Define Testing Requirement and Strategy 103
  - 7.5.2 Prepare Testing Environment 103

cost of implementation by not purchasing some of the modules from an ERP vendor. In the case where stand-alone modules provide satisfactory results to end users, this approach is preferred. For example, Polaris preferred to implement PeopleSoft Human Resource Management System (HRMS) and financial system and integrate it with the project management tool. Likewise, some may perceive SAP's manufacturing tool and customer relationship management (CRM) system to be better than PeopleSoft modules. In such cases, the organization can integrate different modules from different vendors and integrate them.

1.5 Modules of ERP

ERP as an application consists of different modules. Each module typically takes care of one function. Thus, there will be different modules such as: finance asset management; materials management; production management; project management; quality management; maintenance management, sales and distribution; HR management; CRM, etc. These modules usually cater to one function or department of an organization. The ERPs can also have different packages for different industries. These packages are meant for providing solutions for specific industries alone, such as process industry, gas, steel, automobile, textile, cement, banking, finance, etc. In all these packages, the functional modules take care of one function only. However, these functional modules can be integrated later on depending on the scope of the implementation.

In the following section, some of the functional modules have been described to provide an understanding of the information flow given for automating business processes associated with that function.

1.5.1 Components of Material Management Module

Fig. 1.5 indicates automation of business processes in material management.

These are the typical business processes in the material management (MM) module. The automation of these processes ensures that procurement (purchase) and inventory management have seamless information flow. Different activities in MM procurement are:

1. Procurement of materials and services that go for further processing and value addition
2. Identification and selection of vendors
3. Controlling quality and quantity of material and service from the vendor
4. Maintaining a healthy relationship with vendors in terms of payment and other contractual obligations
5. Flow of information at the required level related to procurement activities

Inverting Management		Purchasing	
	Material Master	Logistics Invoice Verification	
Physical Inventory		MRP	Valuation
Service Entry Sheet		Service Master	
Product Catalogue		Foreign Trade/ Customer	

Fig. 1.5 Automation of business processes in material management

While automating these processes, roles and responsibilities of each process are defined. This ensures proper authorization for each business transaction.

The integration of this module to the finance module is required to transfer data for cost accounting (also known as controlling) which happens through the allocation of a purchase order to the cost centre. For a purchase order, a cost centre would require materials and services for procuring and so the cost of procurement needs to be assigned to the cost centre. Through the interface between controlling modules and MM modules, this information is passed, enabling a better visibility and maintaining of business transaction.

The procurement module also interfaces with the accounting module for retrieving a vendor master list. This list maintains details of approved vendors and represents an account called the creditor account in accounting modules. During any purchasing transaction, the accounts can be specified to which the costs can be charged. Similarly, an interface is also credited between sales modules and MM modules, where if any materials are requisitioned, they can be charged to a sales vendor.

The above example shows how each business transaction is interlinked to others across the modules as well as inside the modules. In the case of MM modules, a procurement transaction has to interact with other modules to retrieve data as well as provide information to all stakeholders responsible for this transaction. ERP helps to automate all these transactions with assurance of connectedness of data, ensuring that information is transparent and available to all persons (or stakeholders) responsible for these transactions. Figure 1.6 shows the cycle process of the procurement business transaction to the final transaction.

1.6 ERP Life Cycle

An ERP project typically goes through the life cycle stages as shown in Fig. 1.7.

At the outset, the customer decides to implement ERP so that specific business benefits can be obtained. These business benefits will freeze the scope of imple-

Business process automation can enable an organization to modify or introduce a completely new set of business processes which would transform the way business works. A new business module equipped with better processes to handle customer requirement can average which will become the differentiating factor for the organization. For example, Amazon.com and Dell.com have introduced new business modules of ordering products through the Internet; books, music and any other available products can be ordered from Amazon.com without even travelling to physical stores. Similarly, one can decide and configure a computer or notebook without visiting the store. Expert advice is available on Dell.com which enables the consumer to configure the products according to his/her need. These two are examples of two new business modules which could be introduced because of automation of business processes.

There are also other examples of new business modules being created through automation, including Dena Bank and ICICI Bank created an online banking system which revolutionized retail banking. HLL (now ULL) used process automaton to achieve a zero-stock production system. Thus, automation of business processes has become quite important in corporate strategy. Automation processes' analyses are made on the existing processes, which could result in a further improvement in the productivity. For example, if customer care service is to be automated then first an analysis is made on the response time taken to customer query, the number of activities required to complete a customer query, how many employees are involved in attending the query and the total cost involved in the entire process. This kind of analysis of an existing business area for further improvement is advisable so that the entire process becomes more effective and efficient.

1.8 Benefits of Automation Through ERP

1.8.1 *Collaboration Approach*

The process of doing business and the thought process behind developing new business modules are going through a radical change. Earlier, confidentiality of information and proprietary knowledge were the differentiating knowledge factors. Only a handful of employees used to have access to this confidential and proprietary information, which created a knowledge barrier in the organization. Information flow and knowledge were sacred and this had become an approach for operational management for many organizations. But with changing times, a need to save knowledge and information both inside and outside have become the strategy for thriving organizations. As the enterprises look outwards to take benefit of global economy, the day-to-day activities as well as transactions have became digital. There is a growing need to reduce the cost of operations while being productive. The resultant of these two worlds make a business model viable and sustainable as this makes the enterprise following this model competitive. This can be termed as collaboration approach and ERP helps in achieving this.

1.8 Benefits of Automation Through ERP 13

Through collaboration, employees at different levels share knowledge and best practices; the best practice ensures that the wheel need not be invented again. A business unit can learn from another business unit in terms of best practices, usage and improvement of business processes, usage of technology to deliver products and services to the customer. The capabilities to collaborate and ability to increase the frequency of collaboration and sharing the available knowledge improve performance and productivity. The organization becomes a learning organization, where new ideas are encouraged through collaboration. This collaboration should be done not only internally but also externally with suppliers, partners and all other stakeholders. ERP, through the automation of business processes, helps an organization to share knowledge and best practices amongst its employees. This is done through an employee portal and a knowledge management portal. Stakeholders outside the organization can also access information and would feel part of the progress made by the organization. ERP becomes a common binding face for the organization.

ERP helps to embrace this collaborative approach; integration of different modules helps to retrieve data from a central place and deliver it in the form of online reports. Transparency in all the transactions helps all the stakeholders to undergo a mind shift. Organization structures change giving rise to a flat structure with an attitude to accept open culture. An open culture implies an open attitude towards every individual and partner with whom the employees transact daily. This openness is a result of information exchange between employees, business partners, customers and vendors, which has been made easier and affordable by ERP.

1.8.2 Cross Functional Barrier is Broken

Earlier, different departments in an organization coexisted with walls between them. The data and privacy not only created a syndrome called 'that's not my area' but also created compulsion in the customer's mind. A customer sees the service-providing organization as one and not with different barriers. He expects the best service to his problems and does not care about which department should solve the problem. If an organization has to provide the best solution to a customer and wants to stand out among competitions, then it (organization) has to deliver the solution at a fast speed without caring for the 'that's not my area' syndrome. ERP helps to fulfil this approach by defining roles and responsibilities, by defining a business process flow for each request and by helping to control and monitor each and every request electronically. This helps to increase the effectiveness of the whole organization.

1.8.3 Integration of Island of Automation

In earlier years, each functional area only through about business processes related to its own department only. The automation, if done, was carried out only for the system exclusively in that functional area only. For example, purchase management, inventory management or HR management were the stand-alone subsystems (if an organization is considered as one system, each of these becomes a subsystem). Thus, the island of automation was created.

With collaborative approach, these islands were integrated; the problem with the island approach was that while each function had a robust process and system, this was not accessible to others in the organization. The information available to only select groups of manpower and back office professionals was not available for 'reuse' by other departments. As a result, the resources for data storage, data entry and processing were duplicated. For example, pay roll processing and HR management used personal data for the same employees. But being the stand-alone system, both these data were entered into two different databases by two different sets of data entry operators. This resulted in double investment for IT resources, with the integration of islands not only was the same database available for their entire organization, similar business transactions were not duplicated any more.

1.8.4 Key Principles for Integration of Islands

For achieving an integrated organization-wide system, there should be a collaboration amongst the departments; each department should organize its processes keeping the entire organization in mind. To support this approach, there are key principles that need to be followed:

- Central database
- Role-based system
- Automation of all business processes

Central database helps to reduce wastage of time spent for recording the same data on multiple occasions and in multiple places. The idea is that data and information that are collected by an organization should be recorded once only and also should be stored at one place. This principle ensures that the entire organization has access to the same information all the time. The argument against this principle was that storing the entire organization data at one place would require investment in a high-capacity server. To maintain this server, the skill set requirement is quite different and needs specialized training. However, in modern times this argument no longer holds good and with the progress made in the ability to scale up, database storage space with virtual server technology has been possible at a lower price. This ability along with the capability to process information faster with dual core technology at a lower cost has completely nullified this argument. As a result, most of the business houses are able to expand their wealth of information at a central place without worrying about freeing up storage space.

Central database also helps to digitize information. Paper or hard copies are still preferred by many end users as it is the convenient form of mobile data carrier. But the disadvantage is that at any given point of time, there could be multiple versions of the same information available on paper and this becomes difficult to control. This leads to incorrect decision-making processes and could be dangerous at times. There can also be information security risk when this hard copy falls into wrong hands. With the central database system, information can be retrieved from one place and can be linked (hyperlinked) to other sets of data. This makes it easier for designing the information management system around these hyperlinked data. As a result, with respect to a particular business transaction all possible data are available and can be retrieved at any desired level of details. Even using metadata, keywords can be used for searching information at any point of time. Metadata allow access to all documents having relevant information and save time in searching for information that can then be reviewed and analysed by employees and other relevant stakeholders.

Role-based access helps in securing information at the correct level at anytime, anywhere. With web-based technology, the ability to retrieve information anywhere anytime has become possible. In an organization, based on the role played by an individual, he or she needs to access required information. The management in an organization can decide the amount of information that is necessary for that particular role. This means there has to be a restriction on the details of information needed by him. This restriction can be imposed consistently by the ERP application by using role-based organization structure and by defining security levels, ERP is able to restrict the information level while ensuring that the required information to carry out the responsibilities is available. For example, some information is available in the public domain in the organization website and is assigned the lowest level of security (say zero). The next level (say one) can be assigned to the customer, who can access the organization's customer portal. So all the stakeholders with the role of a customer can access this portal meant for customers only. Similarly, there are different levels of security for all the employees, different level for managers and finally all financial and confidential numbers available only for senior management. Each role is given an ID and password and by using this combination, the employees playing that role can access the required information. With the advancement of technology, biometric security system can be used for authenticating the person.

Role-based access also can be possible with a single sign on, i.e., an employee does not have to login to ERP several times, with initial login to organization network. The employee is able to login to ERP, thus reducing irritation of logging in several times. ERP also allows an employee to play multiple roles. For example, a customer care executive has to answer queries from the customer and provide them with a solution. To provide solutions, the executive has to retrieve technical information from the existing knowledge bank and provide the same to the customer. This dual role is a need for many organizations like Bharati; ERP helps in establishing roles and responsibilities and levels of information required for carrying out these roles. Even ERP allows delegation and extension of roles to other employees.

Automation of Maximum possible business process should be done using flow diagram of business process a owner and assigning appropriate roles for each business process as a owner, each activity showed be completed. This assignment and the flow diagram help in automating the daily activities. This automation helps to make common groups of request, both from customer and employees, which further leads to approval and completion of the activities. This is called word flow

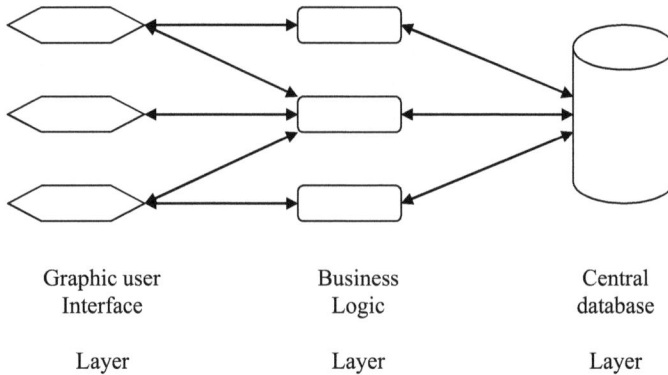

Fig. 1.8 Three-tier architecture for ERP

management, where each activity is assigned a role and another role is authorized to review and complete these activities. This makes faster and effective completion of business transaction.

The above-mentioned three key principles in ERP can be possible by a three-tier architecture. Figure 1.8 shows the required architecture for ERP.

This architecture helps to manage the system centrally. Maintenance can be easier as the central database layer helps to reduce cost and effect required for the distributed database system. Business layer is also located in one layer, which helps to upgrade or modify business logic at one place. This architecture has another advantage that in case the management decide to outsource maintenance of the ERP, then the transition becomes easier.

Document Management System Another advantage of ERP is the creation of a document for statutory and business purposes. Even though ERP enables a paperless organization, it still requires documents for legal authorities, for audit purposes as well as for daily business activities (e.g., purchase order generation, invoicing, goods receipt note, etc.). An organization needs these documents to be generated in a user-friendly manner, without putting extra effort for generating the same. Also as per the need end users should be able to modify these documents with ease (these end users need to have the proper authority as per their role to be able to modify these documents). ERP helps not only to manage these documents but also to create and store them as per the business needs. ERP allows attaching different documents to different roles. For example, documents can be linked to customers, employees, products or projects and these can be made available on a demand basis by the users of ERP. This system thus becomes a medium for exchange and information can be searched using key terms and metadata. It also allows achieving information and retrieving it whenever required.

The document management system can also be made compliant to different qualities system, such as CMMI, ISO, BS7799, etc., which would reduce extra investment that the management would have spent to make their business processes compliant to these quality models.

1.9 ERP Products

As shown in the earlier section, there are a number of products in the ERP market. A few products that are widely used are Microsoft dynamics, Oracle, SAP, Siebal and PeopleSoft. PeopleSoft has been taken over by Oracle; nevertheless, there are a number of organizations that use PeopleSoft as these have been installed earlier.

1.9.1 SAP

SAP was founded as system analyse and programme wick lung by ex-IBM employees in 1972. The acronym was later changed to stand for System Application and Products in Data Processing. In 1976, the company was named a 'SAP GMBH' and in 2005 the name of the company was changed to SAP AG.

SAP R/I was the first internationally known product from the organization. This was launched in 1972. The next version, known as SAP R2, was introduced in 1979. In 1992, SAP introduced a client–server architecture-based product named R3. This was a shift from earlier products which were mainframe-based architecture. The next version of SAP was Internet-based mySAP.com which allowed the users of SAP to login to the system from anywhere. SAP NetWeaver is the next product that came out from SAP, it was platform independent and used service-oriented architecture (SOA). This allowed customized development and also helped to interface with other applications. This was possible as NetWeaver was built using open standards and other industry standards that can work with different technology from different manufacturers, viz., Microsoft, Sun Microsystems and IBM. Thus, while using NetWeaver, the customers do not have to change hardware nor have to purchase any software for customers and the total cost of ownership (TCO) is reduced. SAP NetWeaver has the following modules:

- SAP Web Application Server
- SAP Exchange Infrastructure (called XI)
- SAP Business Information Warehouse
- SAP NetWeaver Mobile
- SAP Master Data Management (called MDM)
- SAP Enterprise Portal
- SAP BI accelerator

SAP operates in many countries which are divided into three different geographic regions: (1) North America and Latin America; (2) India, Australia, Japan and other parts of Asia and (3) Europe and Africa. It has more than 100 subsidiaries and has research centres in India, Germany, Canada, China, Israel, etc. SAP solutions are also focussed on industry sectors, such as manufacturing industries, process industries, service industry, financial service, public service, etc.

Architecture of SAP R/3: 'R' denotes real-time and '3' denotes three-tier architecture. This product can be purchased along with four other applications. These five applications make what is known as SAP R/3 business suite. The four applications are: CRM, supply chain management (SCM), supplier relationship management (SRM) and product life cycle management (PLM). Currently, the target for SAP is to tap the small- and medium-sized enterprises (SMEs) sector as the potential is huge. For this sector, new products such as SAP business all in one, SAP business one and SAP business by design have been introduced.

1.9.2 PeopleSoft

PeopleSoft Inc. was incorporated in 1987 in California, USA. In 2003, it acquired J.D. Edwards, and in 2005, PeopleSoft was acquired by Oracle. Even though it ceased to exist as an independent organization, its products are still being sold in the market. PeopleSoft is well known for its client–server architecture-based HRMS, CRM and financial management packages. Even though there are other products from PeopleSoft such as enterprise performance management and student administration software application, PeopleSoft is still widely reputed for its HRMS and financial packages. It was web-based architecture for its latest versions of the product (version 8) this architecture which is patented as PeopleSoft Internet architecture (PIA) helps an organization to run as a web client and all business function can be accessed from anywhere. Only security policy and system set-up functions are required to be performed on a client machine. As a result, for any organization, transition to Internet-based architecture becomes easy, less time consuming and less expenditure on hardware procurement this also removes the need for extra downloads for plug-ins. Another feature of PeopleSoft products is that it runs on its own proprietary development platform known as People Tools. This allows the flexibility for PeopleSoft to become platform independent, which allows its products to be able to operate on Oracle, Microsoft SQL server, Informix, Sybase, Os/400, IBM DB2 and even on HP all base/SQL. Even an organization implanting PeopleSoft can write its own code to customize the packages.

PeopleSoft also offered J. D. Edwards products such as 'world' on the As/400 platform, a 'one world' which was platform as well as database independent, i.e., one world can function on IBM DB2, Microsoft's SQL server and Oracle database. This product was also known for being able to configure based on any type of network at the customer's site. However, products which were targeted at small- and medium-sized companies could not afford the cost of the full ERP package.

1.9.3 Oracle

In California, under the name Software Development Laboratories, Oracle was founded in 1977, is now present in more than 145 countries and employs more than 60,000 employees worldwide. Besides ERP, it has products such as Database Management System (DBMS), CRM and SCM. In 1979, the name of the company changed to Relational Software Inc. (RSI) and later on in 1983 the company was christened Oracle with its headquarters in San Francisco.

E-business suite is offered by Oracle as a corporate solution to an organization. The suite consists of care business ERP, CRM, SCM, Oracle Financials, Oracle HRMS, Oracle Sale (this is as per the version no. 12, released in January 2007). Each product has several modules and each module can be purchased separately. This reduces the cost of procurement and selective implementation. Oracle Financials has the following modules: (1) asset management, (2) cash management, (3) business intelligence for financials, (4) general ledger, (5) payment management (for multiple currency), (6) payables and (7) receivables.

Similarly, Oracle SCM has the following modules: (1) procurement, (2) planning, (3) logistics, (4) manufacturing, (5) order management and (6) supply chain planning and execution. The modules in Oracle HRMS are: (1) core HR, (2) payroll, (3) Oracle time and labour, (4) Oracle learning management, (5) self-service, (6) recruitments and (7) Oracle advanced benefits.

Most of these modules are developed in Java (Oracle JDeveloper) or using Oracle Forms and Oracle Reports. The new drive in Oracle is to develop products such that nonprogrammers can customize the applications.

1.9.4 Microsoft Dynamics

Microsoft Dynamics was originally a Damgaard Data A/S product. Damgaard was a Danish company which merged with Navision Software A/S in 2000 and the new entity was known as Navision A/S. Microsoft Corporation acquired this company in 2002 and this product is known worldwide as Dynamic A/X or Axapta and at present is available in different local languages making it a multilingual product.

Microsoft Dynamics ERP has the following software as its components:

- Dynamics CRM
- Dynamics Axapta
- Dynamics Great Plains
- Dynamics Navision
- Dynamics Solomon
- Retail Management

The product has different tools such as debugger query interface and code analyser which helps the users to make customized development and modification to the supplied product. The environment in which the product is developed in IDE, Morphx and the environment resides in the same client application as that of other application; this allows modification or new development on the same instance of client.

1.9.5 Comparison of ERP Packages

While comparing ERP packages, the tendency is to compare cost of the application vis-à-vis feature provided by each ERP. However, over the years introduction of role of CIO has changed that thinking process. CIOs use the concept of TCO as the basis for comparison. TCO constitutes the following costs at the least:

- Cost of software or licence
- Cost of hardware
- Cost of annual maintenance contract
- Cost of training
- Cost of implementation and customization
- Cost of acquiring special skills

This skill has also forced ERP vendors to develop products, so that total cost experience for customers is optimized. The vendors also have formulated their marketing strategies and promotional campaigns to make customers aware of this kind of experience of overall cost of ownership. In this section, comparison of major ERPs has been done using total ownership cost as the basis. To arrive at the total cost, all the costs associated with the ERP application life cycle are considered.

Typically, an ERP application has three phases in its lifetime: (1) procurement and implementation, (2) go live and (3) maintenance and support. The elements of cost in phase (1) are cost of license and software. Cost of hardware, cost of customization and interfacing with other existing applications in the organization, and cost of implementation at different locations, go live cost includes cost of training, for computer department (administrative training) and end user training. Similarly, cost of annual maintenance contract, cost of enhancement and cost of skilled resources would constitute maintenance and support cost.

Procurement and Implementation Cost

This phase includes the cost of license, software and hardware, cost of entering enterprise-wide data to the new ERP application, cost of customization and interfacing with other applications. In this phase, the ERP software is installed, configured and integrated with hardware and other applications. The initial steps in installation of ERP are important as incorrect and incomplete installation in the initial stages would lead to different software-related problems resulting in loss of valuable time during the later stages of installation. Thus, if software installation is carried out using tools then it helps to keep the implementation on time which meets all the requirements of the customer. During integration with other applications, the implementation team faces different problems as the complexity of the application to be interfaced cannot be predicted accurately. Even the complexity of transfer of parameters between applications and compatibility of business process flow definition are underestimated during the project estimation stage. This leads to project overrun, as a result of which the implementation takes longer time than anticipated. During implementation, lack of integration tools will delay the integration; and to add to that incorrect or incompatible integration tools will make the integration a complicated affair. These complications make it difficult to establish a general project plan which leads to effort deviation from the initial estimation of timeline for implementation. Hence, to compare ERP packages for their ease of implementation the following technical features embedded in the ERP package should be compared:

1.9 ERP Products

1. Availability of installation tool or installation wizard
2. Extent of configuration possible
3. Ease of modelling the business processes
4. Ease of transferring or porting data from the present system to ERP
5. Availability and compatibility of integration tools

Of the four ERP packages that have been discussed, Oracle and PeopleSoft seem to have a number of tools for the implementation phase. Microsoft has an installation wizard that helps in installation, but a lot of manual intervention is required to complete the installation. PeopleSoft has an installation tool that automates important installation processes and database configuration and thus reduces manual intervention. SAP, on the other hand, has a complicated installation process and requires good knowledge of the product; this implies that SAP implementation should be handled by trained personnel, who can manually intervene if required. Oracle, on the other hand, needs skilled personnel who would be able to configure the database as well as make decisions on key installation steps. PeopleSoft can be configured by business processes which mean, given a set of business processes, the standard configuration has been predefined. This makes configuration a relatively easier task and enables best practices available for a set of business processes for the implementing organization. SAP also has tools that are useful for configuration, but solid understanding of these tools is required for implementation. Between Oracle and Microsoft, Oracle provides a number of tools for defining business processes and data flows. As compared to this, Microsoft does not provide access to end users for configuring the application.

PeopleSoft has more than 1,000 business process flows; using best practices, these business process flows have been predefined in the ERP application which can help in faster implementation of ERP. Similarly, workflows in Oracle help to understand their own business process through process diagrams; this allows further simplification of the existing process and also helps to adopt industry best practices. However, SAP does not provide such predefined process models, which restrict the availability of these models beyond its own features. For Microsoft, customized development work needs to be coded out for any change in available process models. For data loading and moving, Oracle has a tool called set-up which automates and simplifies initial set-up of data. This tool throws questions during implementation; by answering these questions, parameters are generated. These parameters help in defining policies related to expenses, chart of account, etc. which help in moving and data loading. If data are available in Excel spreadsheets, then while installing PeopleSoft, these can be moved into the application database through utilities. PeopleSoft provides these utilities that can be used for data loading and moving from the previous system; similarly, SAP has tools that help in porting data into their own application during installation for all the ERP applications. Similarly, if pre-packaged integration, process-oriented integration and web services integrations are concerned, then Oracle PeopleSoft, SAP provide tools and utilities that help in integration with other applications that are concurrently running in the enterprise. Earlier versions of Microsoft ERP had difficulties in integrating with other applications, but the present versions definitely help in integration.

Maintenance and Support Cost

This phase starts after customization is done and the project has 'gone live'. 'Go live' implies that the ERP application has been implemented and the organization has started using the new application. Any activity after this phase is included in the maintenance and support phase. All the activities in this phase are required to keep the application under control. Some of the activities are applying patches, upgrading present functionalities, fixing bugs, taking regular back-up of present data and managing archival of past data, applying security policy and using diagnostic tools to analyse the performance capability of the present system. In the TCO, maintenance cost is a major contributor. In this phase, there are activities that are repetitive and can be automated. The ERP packages that provide utilities that make this automation a possibility would enhance the total ownership experience. These utilities help in providing upgrades and patches to the installed application without any human intervention. This allows the installed application to stay current on releases, leading to high uptime for the application and less downtime and less business disruption. The effectiveness of ERP packages was compared while considering different activities involved in this phase.

Diagnostic and Technical Support

After packages are installed, any issue that is related to packages are dealt by the vendor support team. These support teams for Microsoft, Oracle and Siebel are available on the web and also at toll-free numbers. The knowledge base is available on the Internet and can be accessed by the support team. However, PeopleSoft has gone one step further by providing a diagnostic utility by embedding diagnostic scripts in the package itself. This diagnostic utility allows customers to send actual problems through web services by providing snapshots to the customer support centre of PeopleSoft. The snapshot travels through a secured network thus assuring the customers that confidentiality of data in the screenshot will be maintained. This capability in PeopleSoft helps in resolving issues faster and customers are assured of business continuity as PeopleSoft has a defined service level agreement (SCA) for resolving these issues. Microsoft, SAP, Oracle and Siebel resolve the issues related to implementation through telephone calls, emails and forum discussions. Most of the time, this takes longer to resolve as the availability of customer care representatives and number of simultaneous issues could have an impact on the turnaround time. This route also needs 'log files' (which would contain chronological details of events in the system) to be sent to the customer care department; this could be time consuming as it means it would take time for the implementing team to understand the log file, download it and then send it to the customer care department. By providing online support and making remote login to the customer's server, many of these issues can be resolved faster. However, because of security policies and firewall settings, these network logins from a different network may not be possible and as such is not encouraged by the customers. SAP has knowledge

1.9 ERP Products   23

management portals where all frequently asked questions and possible resolutions are made available, but are not a favourite among the customers. This is so because of the uniqueness of each issue faced by the customer.

Patch Management

For any product, patches have to be applied after implementation. This patch keeps the product updated with new features, improves performance of the installed current product by providing performance tuning, improves security features, etc. Sometimes, applying these patches would take quite an amount of time and normally disrupts production activities. For ERP products, it is important to understand that it is necessary to apply a patch on each installation of ERP as each has a separate configuration. Thus, patch requirement for one installation will be different from the other. SAP and Oracle publish the patches available on their website. On the website, they also publish details on the patch, release which includes date of release, release notes which describe the scope of the patch and also describes the functionalities which the patch would address. However, if it does not mention the configuration for which the patch can be suitable, it becomes difficult for the implementation team to select and then decide the patch to be installed for their installed package. This requires additional skills and in the absence of that, this becomes a nightmarish experience for the implementation team to implement the patch. PeopleSoft, on the other hand, has a utility called 'Change Assistant' which has business intelligence to do a checking of prerequisites and postrequisites and then make a decision as to which patch is to be applied for the configuration that has been implemented. This makes it easier for the customers to get patches for their installed products. Microsoft provides patches that come out once in 7–8 months. The patches are generic in nature and can be applied across all the implementations. The architecture of Microsoft Dynamics is such that the patch can be applied for all the installations without worrying about specific configurations.

Similarly, whenever there is a need for upgrading the current version, each product provides a different experience. An upgrade can happen only when steps needed for the same are carried out. There are certain prerequisites which need to be identified. After the identification of these prerequisites, the implementation team has to go through the steps. It is here that ERP products have different capabilities while automating these steps. SAP upgrade steps are not fully automated and need skilled resources that can identify and differentiate steps needed for typical configuration. The complexity while performing different tasks for upgrading SAP is considered high. PeopleSoft provides a utility called 'Upgrade Assistant', which assists in upgrading the product. This utility has inbuilt scripts which help in upgrading the product. As a result, time spent for upgrading PeopleSoft products is considerably reduced. The number of steps in upgrading this product has come down to five and thus, the upgrading process in PeopleSoft has the shortest life cycle. The average downtime for these upgrades from PeopleSoft is usually less than 1 day. The experience while upgrading Oracle has not been favourable when compared with

PeopleSoft. Microsoft has infrequent upgrades to its products and the automation provided in the upgrades seems to be adequate enough for the practitioners. Also these practitioners have expressed their satisfaction as less number of upgrades has reduced downtime for the customers.

Product Testing

PeopleSoft allows customers to send snapshots of screens through its web interfaces for diagnosis and debugging. Using the same approach, PeopleSoft allows customers to submit test cases, which helps in simultaneous testing of the installed application at the customer site and at the PeopleSoft development centre, thus reducing the testing cycle. The subsequent release of new versions of products or release of upgrades to the customer happens at a faster cycle time compared to other ERP products.

Oracle does database performance test using test scripts written by its development team. As a result, validation of the performance of ERP happens at a development centre in an environment different from that of the customer. The actual performance at the implementation site will vary from that reported in the development centre, which depends on the complexity of the environment present at the customer site. SAP provides these types of performance testing for customers, but is reported to be quite expensive and, hence, not used by many customers. Microsoft Dynamics also provides performance testing for customers for their implementation and is less expensive compared to SAP. Their usability is also better as customers can use their own test scripts for the testing.

Data Archiving

Data are purged when the system has too much old data which slow down the performance. To improve performance, excess data need to be archived and then deleted from the production database. Oracle allows the old data to be deleted from the production database, but if these data are to be restored, then it becomes almost an impossible task for Oracle ERP. However, SAP and PeopleSoft provide the capabilities to delete data (purging data), archiving and restoring these data back into production, if required. In PeopleSoft, the administrator at the customer site can set his own archiving rules or use existing, predefined archival templates for purging and archiving. This helps in complying with regulation requirements; for example, if banking records in India need to be kept for 7 years after a monetary transaction, then the administrator can set the rules for archival accordingly. These features are not available in Microsoft Dynamics.

Total Cost of Ownership

TCO is a concept introduced by Gartner, which shows the cost associated with the purchase of hardware or software or both. Here, the cost includes direct as well as indirect cost. Direct cost is cost associated with purchase and implementation. Indirect cost is cost associated with maintenance, training, downtime, loss of business opportunity during outage, etc. This concept has been used as a strategy by different vendors for marketing their products. Different vendors use different programmes to educate their customers. Oracle's marketing strategy is based on TCO. Its unique selling proposition (USP) is that if a customer buys the entire e-business suite, then extra cost associated with integration with other applications would reduce. According to Oracle, this would substantially reduce the cost of ownership over a period of time. This period is dependent on the number of e-suite products that the customer has decided to implement. The more the number of products purchased from Oracle, the better is the integration and lower is the cost. Oracle conducts a number of awareness programmes where case studies related to total ownership experience are presented and ROI is shown for Oracle e-business suites. Through these case studies, reduction of cost of integration is highlighted for e-business suite. Also benefits related to reduction in implementation time, reduction in integration cost and finally advantage of vanilla implementation (vanilla means no customization to the base ERP product) are highlighted. Even though many experts question the advantages of vanilla implementation, some of the customers still feel that the TCO experience with its complete e-business suite is better than its competitors.

In contrast, Microsoft has not been so aggressive in defining its total ownership experience. Microsoft Dynamics' strategy has been to reduce total ownership cost by reducing the price of software. Microsoft also has the strategy to reduce total ownership cost by allowing customization with its ERP product. Since customization and integration have been done by the customer, there is no need for providing customized training, which reduces the cost of ownership.

PeopleSoft has a structured, formal training, where they explain through examples how the TCO has been reduced. PeopleSoft has a development centre where the objective is to bring in features that would reduce TCO. The features built into application in this centre reduce implementation time and debugging downtime. On the whole, this reduces TCO.

SAP, on the other hand, is considered to be complex. As far as total ownership is concerned, the cost is considered to be high. The complexity involved in implementation, level of skill required in the resources to carry out customization and effort required for maintenance have resulted in cost overrun and schedule overrun, and consequently has increased TCO. SAP is considered to be the best in terms of number of features, availability of industry specific solution and best practices, but it is not in a favourable position with respect to TCO.

1.10 Trend in ERP: Open-Source ERP

The recent trend in ERP is to move towards open source. Open-source ERP helps an organization that has the capability to use open source for developing its own ERP as the existing products do not meet requirements of the organization. This open-source ERP also helps when the business processes in the organization change rapidly. Using available open source, the organization can change the functionalities of ERP to meet the needs of the changing business processes. Small- and medium-sized organizations, which otherwise cannot buy ERP from the market, as they are quite expensive, can use available resources in their organizations to develop these ERP for their needs at a lower cost.

1.10.1 Definition of Open Source

A software is called an open source (this definition is as per the definition provided by the Open Source Initiative (OSI), www.opensource.org), if the software is available with following conditions

- Free distribution of software, without any kind of cost attached to it
- The source code should be in readable form.
- If someone uses this code and develops it further, then the derived code should be available under the same licence.
- Complies with author's integrity requirements
- There is no discrimination against any group or persons.
- There is compliance to rights for distribution of licence.
- There is no discrimination against fields of endeavour.
- Compliance to the rights for licences for product, licences for restricting other software and licences for neutral technology

The software that complies with all the points is certified by OSI and this software is then considered to be safe for usage as the development cannot be abandoned easily which helps in adopting it for organizations for their own use.

1.10.2 Advantages of Using Open-Source ERP

In any ERP implementation, integration of available systems and ERP is complex and time consuming and requires a correct set of skilled resources for avoiding errors in the later stages. At the end, standard ERP may not meet the business workflow and business processes in totality. In fact, to provide 100% functionality fit for any business, the customization cost in any ERP can be quite high. But if an organization requires 100% functionality fitment in ERP, then flexibility should be available in ERP to do so. This is where open-source ERP fits in. Apart from flexibility and customization, the implementing organization can build add-on modules and interfaces. These add-ons would help to meet the customization requirement from the implementation point of view by editing metadata in these add-ons and also by changing the code. If high-level customization is required (meaning customization at workflow level), then metadata need to be edited whereas for changing daily business transactions, code needs to be customized. Using metadata, the organization can generate forms, data structures and workflow between forms. These metadata are usually stored in eXtensible Markup Language (XML).

1.10 Trend in ERP: Open-Source ERP

Flexibility is also available in terms of upgrades. Since the framework for metadata and custom code are defined by OSI, it is easier to provide upgrades without affecting customization. The open-source ERP also provides support for different languages and regulations for different countries. This makes it easier for adaptation. This makes it an international software and is possible because of regulations laid down by OSI.

Open-source ERP also takes care of security requirements for protecting information at different levels. Depending on the roles, access to information is defined. However, the problem lies with the fact the implementing organization has to define similar roles of its own.

It (open-source ERP) also allows one to become independent of platforms, which allows the package to be run on any operating system. The database is also independent which allows the ERP to be scalable. However, by making the product database independent, some of the features available in the relational database are sacrificed. This is so because by supporting all databases, only common minimal features are possible to be supported. The usual programming languages used for open sources are open-source scripting languages such as Python, Perl, Java, etc. Python has good features such as syntax is concise, it has built-in capabilities for refactoring and is easily readable. Perl is a systematic language which requires a disciplined approach from the developers. Java, on the other hand, has a number of utilities that help while coding. This is widely used and has a strong support team from its manufacturers. Thus, the requirement for skilled manpower as required in standard ERP products is not necessary. This reduces TCO substantially.

In Table 1.1, a comparative statement is shown for different features available for open-source ERPs. The table does not reflect entire features available for ERP, but an attempt has been made to show comparisons between different open-source ERPs based on individual experience.

1.10.3 Conclusion

Open-source ERP allows customization which helps in automating business processes. This customization can become a competitive advantage for the organization adopting open-source ERPs and can differentiate itself from other organizations. Since source code is available along with application development framework, add-ons can be developed for specific business processes. This can be done across different functional areas that can improve productivity as well as reduce lead time for meeting customer needs. A knowledge management portal is developed which can provide knowledge for improving processes continuously that can provide solutions for changing requirements. With open-source code, the organization can change the workflow in the ERP to meet these improvements in the processes at a lesser cost. Thus, the open-source ERP allows automation of changing business processes ttthrough flexibility, platform independence, allows generation on internal knowledge management at a lesser cost.

Table 1.1 Comparison between open-source ERPs

Available open-source ERPs

Features	SQL Ledger	LX Office	TinyERP	GNUe	ERP5	Opentaps	Compiere
E-commerce	No	Yes	Yes	No	Yes	Yes	Yes
Accounting	Yes	Yes	Yes	No	Yes	Yes	Yes
MRP	No	No	Yes	No	Yes	Yes	No
POS	No	No	Yes	No	Yes	Yes	No
Inventory and Warehouse	Yes	Yes	Yes	No	Yes	Yes	Yes
Upgrade and customization	Yes	Yes	Yes	Yes	Yes	Yes	Yes
Architecture	3 tier, web based	3 tier, web based	3 tier client–server		3 tier, web based	3 tier, web based	3 tier client–server
Interfaces	CGI, SOAP	CGI	XML-RPC, Office	XML-RPC, Corba, LDAP	XML-RPC, SOAP, XML	SOAP, CSV,XML	CSV
Programming languages	Perl	Perl	Python	Python	Python	Java	Java
Stability as per OSI conditions	Good	Good	Good	Not yet proven	Good	Good	Good

1.11 MIS and ERP

Both management information system (MIS) and ERP focus on monitoring business processes, data collection and transfer and sharing and publishing information across the organization. Through MIS, the organization collects data from each and every business transaction and then analyses it. After analysis, the resulting information is created which will help in controlling operational activities. This information is used by senior managers to review and exercise management control and make decisions to align the results with business goals. An ERP system integrates different processes and systems and helps to collect data with a high degree of accuracy and integrity. Data entered at one place become available to the entire organization. Thus, the decision-making process in MIS uses the same information throughout the organization bringing in consistency and predictability. The role of MIS is to generate the required report. This is facilitated by ERP with accuracy and at regular intervals.

MIS is used to find out the variances from the planned performance. It is also used to answer what-if questions asked by the management such as what if the sales volume fall by a certain percentage, what would be the resultant impact on the profitability, etc. These types of decision-making processes are possible because of the reports that MIS generates. ERP also pitches in this context. It helps to generate these reports through an automated and integrated system that enables the management for exercising management control. Thus, ERP acts as an enabler for the business while MIS provides the final report for the management decision process. ERP and MIS can complement each other if the systems are well integrated and aligned with business needs. ERP will enable the management to make decisions such as whether to reduce staff, or to go for investment in certain product areas. Because of ERP, inventory management becomes easier with less investment in work in progress. The lead time also reduces for procurement and the development cycle gets shorter. Customer needs can be captured and transmitted faster making the organization more customer responsive.

ERP enhances financial management and increases transparency. This also helps in better corporate governance. Because of its inherent security system, information is available to particular roles only. As a result, using ERP, MIS can be generated for different roles and their performance can be gauged. In older days, MIS required lots of paper movement as reports required to be printed for publishing information to all concerned. However, with ERP, organizations can generate paperless MIS which can be published instantly to all the relevant stakeholders. Figure 1.9 shows the interaction points of ERP on MIS.

To sum it up, we can show the relationship between ERP and MIS to be complementary to each other. ERP helps in data collection and analysis so that MIS can provide the required reports.

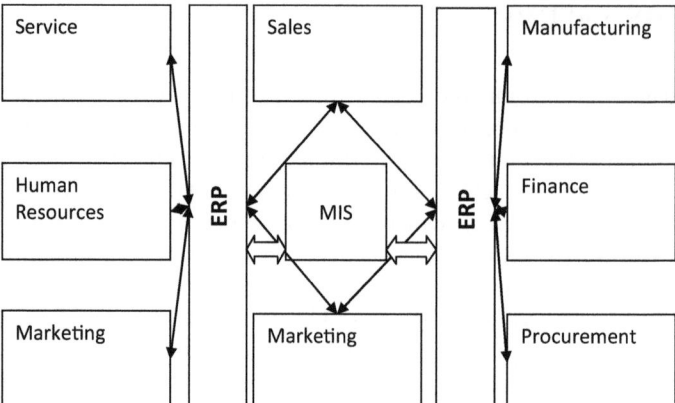

Fig. 1.9 ERP and MIS interaction points

1.12 Summary

As the customer becomes more and more demanding, there is a need to meet the changing scenario with speed and efficiency. While there is a need to increase productivity, there is also a need to reduce cost of operation. The repetitive business processes can be handled effectively by automating them and freeing HR for meeting other uncertainties. These automations not only should be done for each department but also should cut across different departments. Thus, there is a need for automating business processes at the enterprise level. This enterprise level automation started with MRP, then MRP-II, ERP and then finally open-source ERP have taken centre stage. Out of the standard products available in the market, an organization can choose an ERP product for implementation, depending on the features available and the TCO. This comparison helps an organization to choose the product that best suits the needs for the organization.

1.13 Glossary

MRP-II   This is a concept where strategic planning, tactical planning, shop from operation and market forecasting are integrated to bring accuracy in forecasting

ERP   Enterprise resource planning is a tool that helps in planning, scheduling and integrates suppliers and customers as well as in the planning process for an organization. With the help of this tool, accuracy can be obtained for demand forecasting

Stakeholder   A stakeholder for an organization is an entity (it could be a person, organization or any such party) which influences or is affected by the organization's activities

Productivity   In simple terms, it is defined as output divided by input. For any organization, it is important to improve its productivity by optimum utilization of the resources so as to get the maximum output

1.14 Review Questions

1. What is the need for business process automation at the enterprise level?
2. Differentiate between MRP-I and MRP-II?
3. What was the need for ERP?
4. What are the advantages of implementing ERP? How does an organization increase its productivity by implementing ERP?
5. What do you understand by 'cross functional barrier'? Give an example to illustrate the usage of ERP to overcome this barrier.
6. What are the security features available in ERP?
7. What is the advantage of three-tier architecture in ERP?
8. What is meant by TCO? Why is it important to consider this concept while deciding an ERP to buy?
9. Where would you recommend open-source ERP to be implemented? Compare the benefits of open-source ERP with that of standard ERP products.

1.15 Project Work

1. Form a group of five persons. Visit a private sector organization where the annual turnover is less than Rs. 125 crores. Through interviews and discussions, understand the structure of the organization and different departments available. Draw a diagram which would illustrate business processes followed in each department.

    a. Make a decision as to if the organization should implement an ERP?
    b. If the decision to implement ERP is taken, then what would be the break-even period for the investment?
    c. What would be the ROI?
    d. What would be the risks associated with this implementation and what would be the mitigation plan?

2. The same group again should visit a public sector undertaking where ERP has been implemented. Find out if
   a. Objectives of implementing ERP have been met
   b. If the public sector organization has become more productive and has increased its market shares in the market that it is operating, also find ROI for the ERP
   c. If there was a schedule overrun for implementing ERP and reasons for that; what are the lessons learnt from such implementation?
3. Between the two organizations, mark the differences in approach for implementation, differences in ROI calculation as well as percentage improvement in productivity after implementation and prepare a report on the findings.

1.16 Case Study

The situation depicted here is that of a real life situation and the students are expected to understand the situation and come up with holistic solutions and not restricted to information technology alone. The case illustrates that ERP cannot be a panacea for all the problems.

1.16.1 Background

RPG is a leading player in the audio industry. Over a period of 30 years, it has developed a strong partnership with the music industry and has become a force to be reckoned within India. The US$ 25 million organization has its presence in almost all the states in India. It sells cassettes and CDs in different packages and these packages are grouped in different categories, such as movie, artist and situation-based collections.

With so many different product lines, the company wants to simplify its manufacturing operations. It is also looking at improving and upgrading supply chain management systems. At present, it uses its own home-made warehouse management systems in different parts of the company. But the system is less than effective as is evident from the delay in the decision-making process. The delay is primarily in terms of forecasting the sales figures for the next month. The company's system is an integration of systems comprising manufacturing, warehousing, inventory control, planning, shipping and logistics. Apart from these internal systems, the company has to interact with suppliers, musicians and other partners while taking care of both retail and wholesale customers.

Increasingly, the managing director Mr. Goenka finds it difficult to find the numbers related to monthly and annual sales. He wants to build a strong customer base with quality products while keeping his partners happy. He wants to use the feedback mechanism to get information on the quality of cassettes being sold while improving the research activities to keep the company abreast with manufacturing technology. Most of the times, the partners provide information on technology and he needs to make judicious and timely investments to get that 'cutting edge'. His idea is to be the leading music selling company in India at the national and regional level and he aspires to be a recognised player in the international market in 'due course of time'.

1.16.2 Current Situation

When Mr. Goenka matriculated from his Master of Business Administration (MBA) course, he was brimming with ideas and wanted to introduce these ideas in the form of initiatives into his father's company. He felt that these initiatives would launch the company onto a new level that would mark the entry of the company to the international level. However, he had no clue as to what is the present condition of the company.

He divided the company into four regional areas and in each region, subdivisions were made based on the projected sales volumes. But he was not able to give a correct projection of sales figures for each of these divisions and every month, there was a wide discrepancy in the sales forecast versus actual sales. He looked at the models that he had learnt in MBA school for forecasting; each forecasting model used past data to arrive at the forecasting and he had used the past data to derive sales forecasting! He went back to actual sales figures again and he found that actual sales figures as recorded in the books are different from that he had used for forecasting. When contacted, the accounts department said 'this is the latest figure available from sales and these data are based on actual realization of money'. His problem was compounded when within 2 hours a stock-out situation in Coimbatore was reported, whereas Cuttack refused to take any more inventory saying that the Cuttack warehouse was over stocked. One of the suppliers also informed him that because of winter, he will not be able to supply the magnetic tapes in November and December and so wanted to supply excess quantity now to tide over the manufacturing problems later on.

The sales department reported a reduction in sales to the tune of 5% in the quarter and net margin also showed erosion. The competitor T-series on the other hand had a good quarter and increased the volume of sales by 17%—most of the increase coming in at the cost of RPG.

He called for a meeting with regional managers (RMs) and asked them to present their issues, sales figures and projections in the meet. Most of them were not prepared to share the information as they noticed that some suppliers were present in the meet. They sulked and said they would send them by courier as that was the preferred means of sending all information related to sales and inventory. 'That takes hardly 3 days and we can definitely wait for that period'. The sales figures usually arrive from different stockists within 7–8 days after sales and then the RMs compile them for 2 days and courier them to the head office.

1.16.3 Challenges

Mr. Goenka wanted to make RPG follow the model of Walmart and become a true value supplier to customers. He wanted RPG products to be affordable with high-quality music. He also wanted to increase sales figures with reduction in cost while at the same time maintaining a healthy relationship with partners. In fact, he wanted to blur the relationship between suppliers and the company to achieve the same end: 'To sell as many products as possible without both of us having to maintain too much inventory. I also want to be linked to each and every customer so that we can react to their needs at the earliest'.

Mr. Goenka decided to implement ERP and CRM so that he can achieve his vision. He asked all the leading players in ERP and CRM to make presentations to the management team. During the presentation, he found that the team could not answer many of the questions raised by the ERP and CRM vendors. The questions were related to the requirements of the company and the process flow for information. However, Mr. Goenka answered each and every question and a detailed comparative statement was prepared and sent to the board for approval. He was sure that if implemented, this Rs. 1.5 crores (including AMC and hardware cost) implementation will change the way the company works today and will lift the operational efficiency to manifold.

In the next board meeting, his proposal was rejected; he was back to square one.

1.16.4 Discussion Points

1. Is there something wrong in what he learnt in his MBA?
2. Did he make a correct assessment of the present situation?
3. Is there a culture shock?
4. Was he wrong in his recommendations?

1.16.5 Notes that will help in Discussion

1. Past sales data are incomplete and inaccurate; model of forecasting is fine, but without correct data, trend forecast will be inaccurate.
2. Sales data take quite some time to reach the head office, that too in bunches. Since information arrives late, this delays the decision-making process.
3. Lack of transparency as they are not willing to share data with suppliers.
4. Information flow is not defined.
5. ROI on ERP investment was not seen and hence cannot be successful in long-term implementation.
6. The management team including RMs were not with the MD for designing IT strategy and information flow.

Cost advantage: By making correct costing

Differentiation: High quality of cassettes and CDs, controlling costs through PCB

Innovation: Through partners

Growth: Through point of sales tracking

Alliance: Online shopping

Further Reading

Anderegg T (n.d.) MRP/MRPII/ERP/ERM—confusting terms and definitions for a murkey alphabet soup. In: WLUG—Waikato Linux Users Group. http://www.wlug.org.nz/EnterpriseSpeak. Accessed 25 Oct 2007

CIO Magazine's ABCs of ERP

Eckerson WW (1995) Three tier client/server architecture: achieving scalability, performance, and efficiency in client server applications. Open Information Systems 10(1):23–39

Grant D, Hall R, Wailes N, Wright C (2006) The false promise of technological determinism: the case of enterprise resource planning systems. New Technol Work Employ 21(1):2–15

Head S (2005) The new ruthless economy. Work and power in the digital age. Oxford University Press, New York. (ISBN 0–19–517983–8)

Loh TC, Ching LKS (2004) Critical elements for a successful ERP implementation in SMEs. Int J Prod Res 42(17):3433–3455

Monk E, Wagner B (2006) Concepts in enterprise resource planning, 2nd edn. Thomson Course Technology, Boston. (ISBN 0–619–21663–8)

Waldner J-B (1990) Les nouvelles perspectives de la production. Dunod Bordas, Paris. (ISBN 9782040198206)

Waldner J-B (1992) CIM: principles of computer integrated manufacturing. Wiley, Chichester. (p 47; ISBN 047193450X)

www.ittoolbox.com (2005)

www.opensource.org (2005)

Chapter 2
Introduction

The Enterprise Resource Planning (ERP) application implementation methodology is composed of well-defined processes that can be managed in several ways to guide you through an application project. This methodology provides the tools needed to effectively and efficiently plan, conduct, and control project steps to successfully implement new business systems.

The implementation methodology defines an organization's business needs at the beginning of the project and maintains their visibility throughout the execution. It defines time-sensitive business events, and maps each event to the corresponding business and system processes. Using this method, the business community gains an accurate understanding of the business requirements to be met by the final system.

This was designed with scalability in mind. From the largest, multinational, multisite, multientity projects to the smallest, limited-size, constrained-scope projects, the implementation methodology provides the scalability that your project demands. It also allows you to tailor your own approach to match your organization's specific needs. This methodology is also flexible and extensible.

The implementation methodology activities are conducted in phases. These phases provide quality and control checkpoints to coordinate project activities that have a common goal. During a project phase, your project team will simultaneously be executing tasks from several procedures.

The tasks against each phase are organized into *processes*. Each process represents a related set of objectives, resource skill requirements, inputs, and deliverable outputs. There are 13 processes in all, divided into seven phases.

Table 2.1 illustrates the phases and the corresponding processes under each phase.

The following are the typical critical success factors (generic in nature) for the success of an ERP application implementation project:

- Sufficient infrastructure
- Clear understanding of business needs
- Upper management support
- Strong program/project management
- Team strength
- Organizational readiness
- Sufficient technical architecture

Table 2.1 Phases and their corresponding processes

Processes / Phases	Definition	Requirement (Requirement)	Solution (Design)	Build	Testing	Production (Go Live)	Rollout
Project Initiation	■						
Core Process Analysis		■					
Conference Room Pilot			■	■			
Customization/Build			■	■			
Interface/Conversion Building			■	■			
System/Integration Testing				■	■		
Training					■		
User Acceptance testing					■		
Production Go-Live						■	
Roll out							■
System Administration	■	■	■	■	■	■	■
Project Management	■	■	■	■	■	■	■
Change Management	■	■	■	■	■	■	■

The preliminary application implementation project estimate is based on:

- What is the business objective?
- What is the scope of the project and how does it relate to the overall objective?
- Who will participate in the project (employees, consultants, or vendors)?
- What are the constraints affecting the project (timing, budget limitations, or organization changes)?
- Which applications will be implemented?
- Which sites will be involved?
- Will a phase deployment be employed? If so, in what sequence will the applications be implemented?
- When will work commence?
- What experience does the organization have regarding the technology that will be used?

Before creating a detailed work plan, develop the project scope, suggested approach, and preliminary budget.

The indicative phase schedules given in the respective phases in this document are based on the assumption of an average complexity of implementation, which includes the following assumptions:

1. Industry standard processes
2. Two to three sets of books, 5/6 segment chart of account (COA) structure, and a single calendar

3. No Multi Currency (MRC) or GCS requirements
4. Simple multiorganizational structure with one business group, 2/3 LE, 2/3 OU, and corresponding inventory organization
5. No localization requirements
6. Single geographical entity
7. Not a material-intensive business
8. Around 1,000 vendors and customers
9. Moderate volumes of transactions (~500 invoices per month)
10. Dedicated business process owners' time

The color codes depicted in the diagram above for phases and processes are maintained throughout the average case phase schedules to identify the processes and activities in their corresponding phases.

2.1 Target Audience

The ERP application implementation methodology is a high-level description of how the implementation methodology can be used to facilitate implementation projects. Project managers will use this handbook for project planning and scheduling. Team members can use this book to gain a broad understanding of end-to-end implementation of ERP application. It also provides detailed information on the process and tasks involved in each phase of an ERP application implementation life cycle.

2.1.1 Multiple Deployment Sites/Multiphase Considerations

One of the prerequisites before planning an ERP application implementation project activities in line with the application implementation methodology and tailored to the unique specifications of the project is that you need to determine if the transition into production needs to occur for all organizations at the same time. If a phased implementation is planned, decide what applications are to be implemented in each phase.

Consider the following factors when deciding on a single-phase or multiple-phase deployment:

- Is the organization facing changes from other sectors? Give time for an organization to absorb major changes before introducing new factors.
- Is the organizational culture amenable to integrate the new system into its operations? The risk is less if the organization is accustomed to using a common system with related policies and procedures. If business units have been operating independently, the organization must adapt to a cultural change, as well as to the new system.

- Can the project team adequately execute the project so that there is minimal risk in implementing all applications for all organizations at one time?
- Can two or more organizations all fit (from a sizing point of view) on the same servers?
- What is the company's experience with previous system implementations?
- What is the project team's experience with previous system implementations?
- Is the application a production that is controlled, or is it a beta release of the software?

To minimize risk, use the pilot implementation technique. Go live with a carefully selected subset of users who are well trained and able to deal with initial problems. Initial success at a pilot site often sets an attitude of success that carries over to the remaining sites.

2.1.2 Structure of the Document

After this, each section of this document describes one significant phase of the life cycle of the project. The phase description contains:

- *Objectives of the relevant phase*
- Activity flow diagram
- Implementation schedule
- Prerequisites
- Detailed activities containing the tasks corresponding to that activity
- Phase deliverables
- Decision matrix/checklist to gauge the successful completion of the phase and a prerequisite to proceed to the next phase
- Critical success factor

This document describes the following 13 phases of implementation:

- Project initiation
- Core process analysis
- Conference room pilot
- Customization design/build
- Interface/conversion design/build
- System integration testing
- Training
- User acceptance support
- Production go live
- Rollout
- Project management
- System administration
- Change management

The formats of templates corresponding to the deliverables of each phase are provided as an attachment with this document.

2.2 Project Initiation

2.2.1 *Objective*

The objective of project initiation is to frame a standard approach to project management to accomplish and execute a successful ERP application implementation project.

The goal of the project management framework is to prepare a unified project plan to provide a framework, in which all types of ERP application projects can be planned, estimated, controlled, and completed in a consistent manner. This consistency is necessary in an environment where projects use a variety of methods, tools, and approaches to satisfy business needs.

The overall organization of a project management plan is expressed as a process-based methodology, which can be tailored to a project's specific needs. The five management processes are:

- *Control and reporting*—contains tasks that help to confirm the scope and approach of the project, manage change, and control risks. It guides toward management of project plans and reports project status.
- *Work management*—contains tasks that help to define, monitor, and direct all work performed on the project. This process also helps to maintain a financial view of the project.
- *Resource management*—provides guidance on achieving the right level of staffing and skills on the project, and on implementing an infrastructure to support the project.
- *Quality management*—provides guidance toward implementing quality measures to ensure that the project meets the client's purpose and expectations throughout the project life cycle.
- *Configuration management*—contains tasks that help to store, organize track, and control all items produced by and delivered to the project. The configuration management process also calls for providing a single location from which all project deliverables are released.

Each of the above processes needs to be framed with a common format of task categories, which include:

- Planning tasks with respect to project scope, quality, time, and cost
- Controlling tasks performed concurrently with execution tasks basically measuring performance and taking corrective action as needed
- Completing tasks to formalize acceptance of project deliverables and obtain a proper sign-off of all the deliverables

Figure 2.1 depicts the fundamental values of a project management approach.

The project initiation is the kickoff or the start of a new project with the project manager, and a few key resources that basically form the project management team are required to carry out this phase.

Fig. 2.1 Fundamental values of a project management approach

Before embarking on the creation of a project plan, there are certain golden rules which each member of the project management team needs to keep in mind and practice as necessary. The golden rules are as follows:

- *Start right*—if you start badly, it is always difficult and often impossible to recover.
- *Know your client*—get to know the project stakeholders and, in particular, who is funding the project. These two parties usually are not the same people. Understand their spheres of influence and agree on when and how their support will be called upon, and what the project will expect of them.
- *Define the project scope*—study the contract, reach an agreement with the client on a precise definition of the scope of work and terms of engagement, and understand how a consultant's scope of work relates to the client's business objectives and the key benefits the client expects to achieve from it.
- *Manage the risks*—determine the key risks, analyze their impact, define containment strategies, and establish contingency plans.
- *Field a winning team*—select people based on three primary criteria: merit, ability to adapt rapidly, and personality fit with the team. Set for everyone involved, the expectation of "no surprises." Adopt a conflict minimization strategy and try to reach a win–win solution.
- *Produce formal documentation*—document all deliveries, agreements, decisions, issues, resolutions, actions, and file them with all correspondence and minutes of the meetings.
- *Plan for completion*—*project completion* is the final phase of the change, process brought about by the project. Plan early for it. Final impressions are the ones that you leave with people.
- *Communicate with honesty and conviction*—set standards with the stakeholders and the team on communications and progress reporting, adhere to them, be honest, and do it frequently.

The major activities during a project initiation phase are as follows (Fig. 2.2; Table 2.2):

- Contract review and firm up project scope, assumptions, exclusions, and deliverables
- Prepare a unified project plan

2.2 Project Initiation

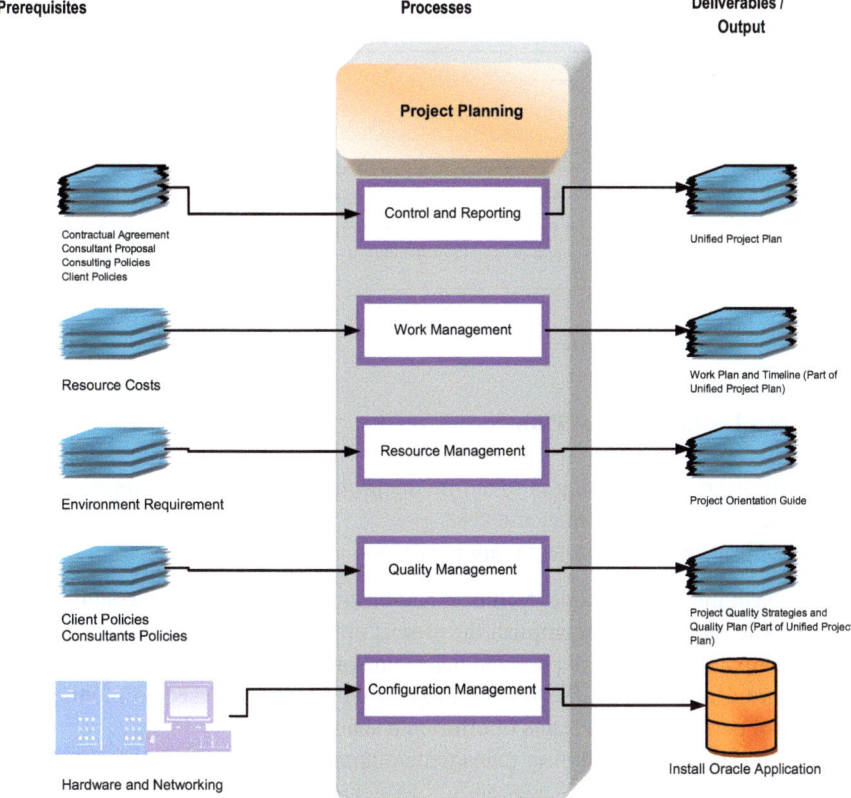

Fig. 2.2 Business flow diagram: project initiation

- Field a project team
- Install the software

2.2.2 Prerequisites

- Contractual agreement documents
- Consultant proposal
- Client policies
- Consulting policies
- Resource costs
- Environment requirement

Table 2.2 Implementation schedule: project initiation

Activities	TimeLine							
	Month 1				Month 2			
	1	2	3	4	5	6	7	8
Contract review and firm up Project Scope	■							
Prepare Unified Project Plan	■	■	■					
Field a project team	■	■						
Install the software		■	■					

2.2.3 Detail Activities

Prepare a Unified Project Plan

The goal of project planning is to define the project objectives and approach with respect to scope, quality, time, and cost.

The objectives of the project planning are to:

- Establish the project scope, technical and business objectives, and resources and schedule required to accomplish the project objectives.
- Develop a baseline work plan and determine project resource requirements.
- Prepare a resource profile for the project, which will be used to monitor and control performance in terms of effort and schedule.
- Obtain client and consulting management approval to proceed with the execution of the project.
- Determine the measures, which will be used on the project to measure and maintain the quality of the project processes and deliverables.

The key contents of the unified project plan should contain the approach and details of the project execution and management corresponding to the following:

1. Establish scope, objectives, and approach
2. Define control and reporting strategies, standards, and procedures
3. Establish work plan and detailed activity schedule
4. Define resource management strategies, standards, and procedures
5. Establish staffing and organization plan
6. Create project orientation guide
7. Establish infrastructure plan
8. Define quality management strategies, standards, and procedures
9. Define configuration management strategies, standards, and procedures
10. Define risk management strategy, identify probable risks, and define mitigation plan.

The key techniques to be employed in this process during project planning are *work planning, estimating, risk mitigation planning, and resource/effort loading profile*. These techniques support the development of the initial project work plan and effort profile.

2.2 Project Initiation

Table 2.3 Deliverables: project initiation

Sl No.	Name of deliverable	Template ID	Responsibility	
			Major	Assisted
1	Unified project plan		Consulting organization	Client
2	Project orientation guide		Consulting organization	Client
3	Client policies and procedures document		Client	
4	Client requirement documents or statement of work as prepared by the client before signing of contract		Client	

Reuse of deliverables is an important asset to the project since it will reduce the overall preparation duration of the documents. *Project planning must also incorporate the planning for reuse of deliverables by incorporating the time required to prepare the deliverables and inclusion in the knowledge repository* (Table 2.3)

Implement a Project Team

Resource management during project initiation is somewhat unique in a sense that you not only plan but also oversee implementation of human and physical resources at the start of the project. In addition, you not only do long-term planning but also plan a core set of resources needed to support the execution of the first project phase. The primary techniques you use in this process during project planning are organizational design, recruiting, and negotiating.

The basic tasks for the above activity are as follows:

1. *Plan a strong project team*—working with the client project manager to plan out a single project team including consulting and client staff. You will need to negotiate with both, the client to agree on an acceptable level of client resource commitment and with your own management to secure commitments for consulting staff to form your core project team.
2. *Create a project orientation guide*—establish the expectations of the team members early in the engagement, facilitating the creation of a strong team. Also, the very important process of time collection will be restated in the project orientation guide, so that all team members clearly understand the importance and ramifications of time collection and reporting.
3. *Prepare a working environment*—the project staff should have a designated work area at the client site. Make sure that suitable workstations, phones, and office supplies will be available to each full-time project member. Agree with the client on arrangements to accommodate new full- and part-time project staff in the work area. Having to find work areas, each time a new staff member comes on to the site, is a distraction that reduces work productivity.

Table 2.4 Major activities, checkpoints, and weightage

Major activities	Checkpoints	Weightage
Prepare unified project plan	Is the preparation of a unified project plan complete?	
	Is the above plan signed off and accepted by the client?	
	Is a minimum project team to start the core process analysis in place?	
	Is the project plan and activity-wise schedule communicated and accepted by the project team?	
Install ERP application	Is the required hardware and software available?	
	Are proper working environment and facilities available to the project team?	
	Is networking infrastructure available at the client site as a prerequisite for accessing the software application?	

Install ERP Application

The prerequisites for installing the software are:

1. ERP application servers are available.
2. Server specifications are according to the requirement of the project system architecture.
3. Software is available.
4. The software version is according to the requirement.
5. Proper networking arrangements are done.
6. System administrator is available to the project team to carry out the installation.

The key activities include:

1. Preparing for installation—check the above prerequisites and plan for installation like creating privileges, access, storage space, partitioning, etc.
2. Installing ERP application software

Carry out the postinstallation activities like setting up roles, responsibilities, backup tasks, etc.

2.2.4 Decision Matrix/Checklist

Note that this checklist is indicative of assessing whether the activity is completed or not and giving feedback/escalation as deemed necessary. This does not stop in proceeding to the next phase of business requirement definition in any way (Table 2.4).

2.2.5 *Critical Success Factors*

The critical success factors of this activity phase are as follows:
- Scope, objectives, and approach are agreed on and understood by all parties.
- Project culture and climate are established conducive to a win philosophy.
- Risks are identified and containment measures are put in place.
- The client accepts the project work plan in the context of the project's scope and risk assessment.
- The client understands the obligation to provide resources to support the project work plan.

Chapter 3
Core Process Analysis

3.1 Objective

The purpose of the core process analysis is to define the business requirements of the new application system aligned to the future business model as conceived by the client (Fig. 3.1; Table 3.1).

The principle activities during the business requirement definition are as follows:

- Examine and document current processes and practices to understand the main business factors that currently benefit the business.
- Gather business transaction and data volumes from the future business model to help assess the system's ability to support current and future business volume.
- Carefully document audit and control requirements to satisfy financial and quality policies.
- Identify the business-operating requirements that the technical architecture will need to support.
- Analyze and identify the reporting requirements for the business.

Business requirement mapping and gap analysis is an iterative approach with the following objectives:

- Prepare business process designs through mapping with standard functionality within the system.
- Identify gaps in the application.
- Propose feasible bridges to gaps agreeable to the client.
- Freeze the future business process model.

Areas that are necessary to map include:

- Business requirements attached to business process steps
- Report requirements
- Business data requirements

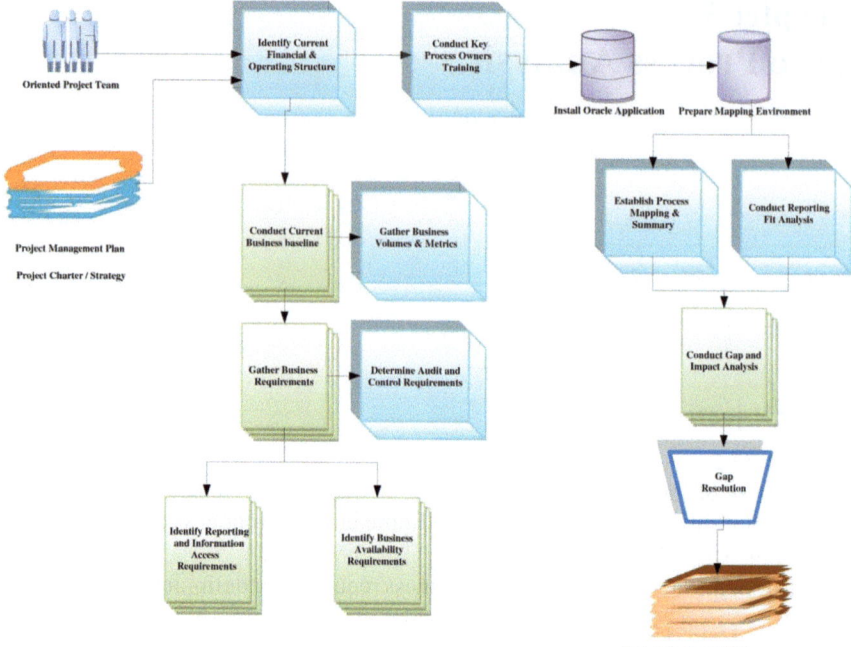

Fig. 3.1 Business flow diagram: core process analysis

3.2 Prerequisites

3.2.1 Business Requirement Definition

- Dedicated resources for conducting analysis
- A committed project sponsor who maintains a consistent and high level of team commitment
- Active involvement and support of management
- Active involvement and support of knowledgeable business area specialists
- Full access to information about relevant business areas, their processes, data generation, and use
- Current business process documents for reference purpose
- Current business reports for reference purpose

3.2.2 Business Requirement Mapping

- Current business baseline document
- Consistency of team composition across process design, mapping, narrative writing, and approval activities

3.3 Detail Activities

Table 3.1 Implementation schedule: core process analysis

Activities	TimeLine												
	Month1				Month2				Month 3				
	1	2	3	4	5	6	7	8	9	10	11	12	13
Current Business Process Analysis	■	■											
Business Transactions and Data Volumes		■	■										
Audit & Control requirements			■	■									
Business Technical Architecture			■	■									
Business Reporting Requirement				■	■								
Install application													
Key process owners training					■	■							
Prepare mapping environment						■							
Perform Business Requirement Mapping							■	■					
Identify gaps in the application								■	■				
Gap Resolution										■	■		
Freeze the Future business process Model												■	

- Consistency of team approach across business areas or process groups
- Clear and concise visualization regarding how information will flow across and be owned by organizations, functions, and applications
- Availability of system and software
- Availability of hardware servers and client machines

3.3 Detail Activities

3.3.1 Business Requirement Definition

Identify Current Financial and Operating Structure

Information about current and proposed organizational structures may have been discussed during the acquisition cycle and may have been one of the deciding factors in the selection of the application package. Be sure to gather this information if it is available.

Start the internal organizational analysis by interviewing the highest-ranking financial official possible, since the holder of that position will be likely to have most knowledge of the financial and operating structure of the organization.

An organization's operating structure drives the business and has a strong influence on the setup and use of the applications. However, the financial statements will reflect the operating structure to allow profitability, balance sheet and cash flow reporting, and analysis against that structure. Reporting and analysis begin by capturing and valuing operational transactions that occur at the specific event level. These events occur at a specific site that must be properly defined and ultimately set up at the correct organizational level and with the following appropriate organizational attributes:

- Interview organization management to obtain a clear understanding of current and proposed entity structure
- Develop a chart showing the current organization structure
- Develop a business organization overview and listing
- Define and chart the financial operating environment
- Define the financial and management reporting environment

Conducting Current Business Baseline

Conduct a baseline to develop a common understanding within the project team and across the organization of what cross-functional processes are in place to support the achievement of business objectives in the current environment. Awareness of current business requirements and unique processes today will educate team members and prepare them for the construction of future business processes.

As many current environments may not have been developed in an integrated fashion, and people may have entered the organization at different times during the evolution of the current business systems, many people and departments may not fully understand the processes and requirements of other departments. An added benefit of baselining is the resulting cross-functional knowledge gained by team members. This information is invaluable during subsequent mapping of future business processes and requirements to new application functionality.

Use structured process questionnaires (*refer current business baseline questions for all module templates*) to collect business and current system information during a business baseline interview for a given process. These questionnaires can be modified to help make sure that the team interviews include the following factors:

- Business events—triggers for action (for example, receive invoice)
- Location, nature, and sequence of transactions—data added
- Magnitude and frequency of transactions
- Performance metrics, core processes, or critical transactions
- Key factors for success
- Key processes and process cost drivers
- Representative families or products and transactions
- Opportunities, constraints, risks, and issues
- Underlying structures of static data organization
- Bottlenecks to the flow of information and material
- The particular value of current business processes
- Data-gathering methods that drive technology requirements
- Current system configuration options
- Schedule, confirm, and prepare for process definition sessions by business area
- Identify the core business processes (level 1 if you are using enterprise resource planning (ERP) business models) and write a summary description of each process
- Conduct interviews using the questionnaires and other sources of information to clarify questions you have identified

3.3 Detail Activities

Table 3.2 Business volumes and metrics

Process Ref No.	Core process analysis Task
1	Review the existing documentation
2	Gather business process measurements
3	Extract business volumes from the current business and the business requirement scenarios
4	Summarize the business transaction volume statistics by functional area
5	Gather total data volume requirements
6	Gather total data volume requirements
7	Determine the critical processing period's window
8	Gather system user counts by functional area

- Gather any other current business materials that may enhance team understanding and documenting of current business process requirements
- Identify any issues regarding the current business analysis

Gathering Business Volumes and Metrics

The business volumes and metrics document the data volumes and processing frequency of the transactions on the new production system (Table 3.2). To begin this task, examine all major business processes that transact moderate to high volumes of data (for example, customer orders, purchase orders, purchase requisitions, manufacturing orders, manufacturing receipts, invoices paid, and journal entries).

Concentrating on the resource-intensive areas allows you to assess when the new system will inherit the same volume and performance challenges. If you are working on a small, single-site implementation, it may seem that there are no performance risks. Do not minimize the importance of this task. The number of possible configurations can lead to performance problems, even for smaller implementations.

Gather Business Requirement Scenarios

The rule is: You must express all business requirements in the context of process models and business requirement scenarios. In other words, you must be able to trace all detailed business requirements to business processes.

A business requirement specification (BRS) is a formal statement of the detailed business requirements for a business process, the source of these requirements, how these requirements will be satisfied (either by the application, manual process steps, workarounds, or by other applications), and what prototyping steps must be taken to prove the designs (Table 3.3).

Since BRS development sessions are design sessions, you can expect additions and corrections to be made to the initial process models and function models during the course of design. Although it is possible to create a BRS at the same time that its process model is being developed, it is actually better to start with a graphical rep-

Table 3.3 Business requirements

Process Ref No.	Core process analysis Task
1	Train all the assigned team members to use the methods and tools for BRS development, and in the boundary and characteristics of the target business process
2	Construct process identification for the target future business process. If it exists, use the preliminary BRS collected during initial project planning
3	For each process step, document business requirements and indicate the source of those requirements

Table 3.4 Audit and control requirements

Process Ref No	Core process analysis Task
1	Review current security, manual operations procedures, and future business requirement scenarios
2	Evaluate audit specifications for division of responsibility in finance and operations
3	Create a list of security requirements for the organization, operating system, application, and database to support future business processes

resentation of the business process before going into descriptive detail since people tend to respond better to pictures than to words. Using visualization techniques during the early stages of any design is crucial for understanding and for the common agreement.

Determine Audit and Control Requirements

The overall objective of audit and control requirements is to consider audits and controls that will reduce or minimize the risk of those transactions being executed that place organization assets or information in jeopardy (Table 3.4). Such transactions, if executed, should be detectable and their recurrence prevented.

It is helpful to think about both application and general controls and risks. The following tables may help frame the thinking process:

Application risk	Application controls
Unauthorized application access	Logical access controls
Incorrect data entry	Input controls
Rejected items resolution	Processing controls
Incorrect processing/reporting	Output controls

3.3 Detail Activities

General risk	General controls
Unauthorized system access	System access controls
Unauthorized program changes	Program change controls
Inadequate information systems operations	Organization controls
Business interruption	Disaster recovery controls

From an auditing standpoint, these are the questions you are attempting to answer:

- What was changed?
- Who changed it?
- When did they change it?
- Why did they change it?
- Who logged in?
- Why did they log in?
- What kind of monitoring takes place for transaction processing?

From the standpoint of financial controls, consider the following questions:

- What steps need to be taken to facilitate external auditors' review of procedures and controls?
- What controls are in place to prevent insider trading, fraud, nonmalicious errors, and so on?
- What approval hierarchies need to be in place?
- What business transactions are permissible in a web-based system?
- What kind of special security is required for financial and confidential information systems (such as general ledger, accounts receivable, human resources, electronic funds transfer, electronic data integration, and data warehouse)?

Identify Business Availability Requirements

What constitutes business availability is often regarded as the sole responsibility of the information systems' organization.

This conventional reasoning does not give adequate representation to the business operations and user communities who may have a different view of what availability is necessary in order to perform critical business functions and provide customer satisfaction.

This task is designed to facilitate a discussion between the operational and information system departments of the business in order to arrive at a consensus on what is an acceptable percentage system uptime and the contingency measures to implement during a system outage. The system architect, database administrator, and system administrator should represent the information systems organization. Key business analysts and lead users should represent the business operations and user communities in the discussion.

Table 3.5 Reporting and information

Process Ref No.	Core process analysis Task
1	Review current-reporting materials that may enhance the team's understanding of the current-state reporting environment
2	Determine an approach for collecting report requirements
3	Update the master report tracking list with information from the current reporting materials
4	Update the master report tracking list with information from the business requirement definition documents
5	Identify critical reporting issues and document them
6	Collect business data elements required to be seen on the screens

Identify Reporting and Information Access Requirements

The master report tracking list is used as the primary repository for all information collected about a report requirement (Table 3.5). It should contain system and report name, business purpose, frequency, priority, user name, and contact information.

Possible methods to determine report requirements include:

- List report on the current system
- Future report requirements from the business requirements scenarios
- User survey: The goal of the survey is to collect business critical report requirements that may have been overlooked and not to list every report that may have been run regardless of whether or not anyone used it. Emphasize to the users that only critical business requirements are important, and that their manager must approve their list before mapping begins. Analyzing, mapping, and building reports are very expensive.

You may use one or a combination of these methods.

Also, collect the relevant business data which the user is required to see on the various windows of the legacy systems and understand the usage of the same.

3.3.2 Core Process Analysis: Key Process Owners Training

The main objective of the key process owners training (Fig. 3.2) is to familiarize them with the standard functionality, basic looks and feel, and the navigation methods of ERP applications. This will help the project team in a meaningful discussion on the mapping and gap analysis/resolution where the key process owners can add value to the issues and close the gaps with meaningful resolutions.

3.3 Detail Activities

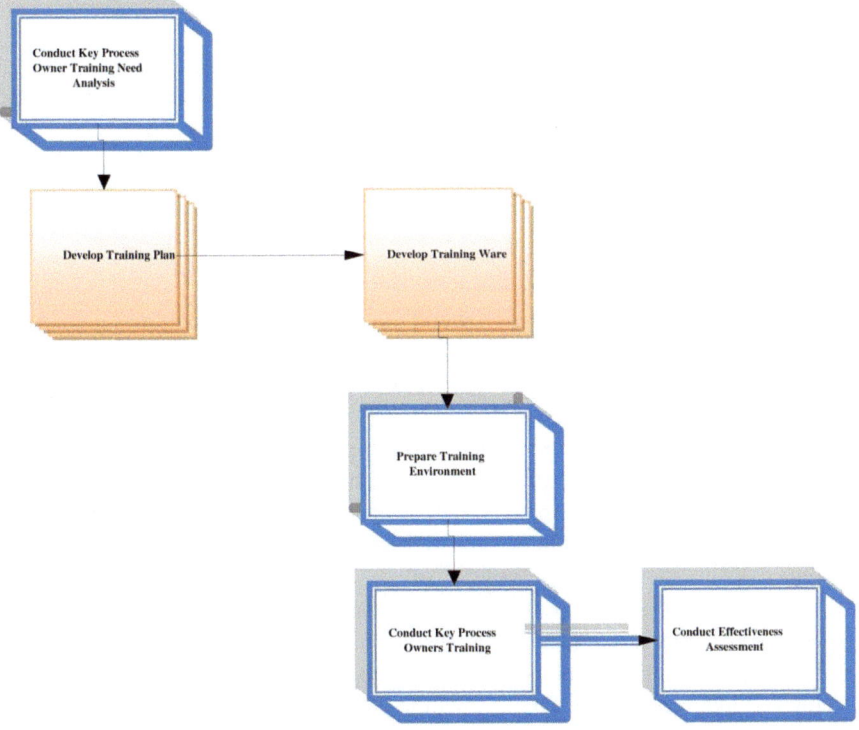

Fig. 3.2 Business flow diagram: core process analysis

Conduct Key Process Owners Training Needs Analysis

During this task, gather information about the knowledge, skills, and aptitudes of all key process owners. Record current knowledge, skills, and aptitudes; compare them to the new competencies; and identify the gaps between the present and the ideal. These gaps mark the parameters of the needed learning. The profile is a snapshot in time which serves not only as a guide for creating learning paths but also as a baseline against which gains can be measured. In addition, the learning needs analysis provides insights into how ready the audience is for the project.

For representation, you must interview a cross section of process owners who are selected by the client, based on their ability to impact the business results anticipated as a result of the implementation. Choose process owners who are empowered to take decisions that affect key business processes of their corresponding areas.

Prepare a Training Plan

- Review and select recommendations from the training needs analysis.
- Create a learning objective and training strategy.

- Tailor learning content for each role.
- Select learning approach and delivery method.
- Describe the approach for creating needed learning materials.
- Determine resource requirements, such as facilities, equipment, materials, and supplies, including learning environment.
- Describe the plan for learning logistics and administration.

Develop Training Material

- In this task, you tailor learning ware and other skills-change materials to meet the learning objectives developed in the user learning plan.
- Focus on developing materials that are user friendly, attractive, and meet good communication standards.
- Develop the training materials per learning path by role, e.g., prepare training material for each module of ERP applications being implemented, etc., for the key process owners. This material will be different in content, look, and feel than the material which needs to be prepared for the end users during end user training.
- Develop the measurement materials like hands-on test cases or exercises, which can measure the extent of training, absorbed by the key process owners.
- Develop the training administration materials—including materials for announcing and logging learning events, tracking participation, and so on.

Prepare Training Environment

- Install user learning environment.
- Set up applications.
- Set up support infrastructure.
- Convert or add necessary sample data.

A separate environment is the preferred choice for learning, because it can contain clean setups and actual data and remain unaffected by concurrent testing and mapping activities.

As part of preparing the user technical learning environment, make sure you test the client desktop devices and other hardware, such as printers.

Key user training often uses the demonstration database shipped with the standard applications. This environment is most appropriate if you plan to use the standard learning materials for user learning, because all examples in the standard notes reference this database.

However, it is not ideal because the data and scenarios will not be familiar to users. So, it is advised to make proper references to the client business processes as documented in the business requirement definition document and give the participants a feel of how these processes will function in the standard ERP application environment. It is always advised that the project team be already oriented with the client business requirement definition when imparting the key user training.

Conduct Key Process Owners Training

The general goal of the user learning events is to conduct and track the skills-change events designed to provide the groups of learners with the skills they need to meet the performance objectives of their new roles. You monitor the pulse and progress of the user learning events as they unfold, to make sure that the momentum and quality are maintained.

As a success factor for the learning events, you develop communications based on the purpose, value, context, and overall logistics. Tailor the messages to the various groups of learners. The communications set the proper tone for the learning, for example, important effort, tailored to the learning styles, providing variety, reinforcement, and so on.

It is extremely important to reinforce and impress upon the client that this training is aimed at creating an understanding of the standard features and standard process flow of ERP applications so that the client can add value to the mapping, gap analysis, and classroom pilot (CRP) sessions to be held jointly with the project team.

Position the learning events in the context of the whole project and the expected business benefits. Ideally, develop a highly interactive campaign to address the changes in roles and performance expectations in a positive and motivational manner:

- Prepare for imparting the training.
- Hold the training events according to the training plan.
- Monitor the performance deployment process.
- Assess the training effectiveness from performance on test cases and participant feedback.

3.3.3 Core Process Analysis: Business Requirement Mapping and Gap Analysis/Resolution

Prepare Mapping Environment

The configured mapping environment prepares an application environment for detailed mapping activities. Try to use an environment that accurately reflects the organization's business to the extent known at that time.

If space and time do not allow for a separate environment for mapping, you may have no other recourse than to use either the training or vision demonstration environment. Be aware of the limitations in using these environments for mapping. The

Table 3.6 Preparation for mapping

Process Ref No.	Core process analysis Task
1	Review architecture requirements and strategy to understand the strategy for deployment of project environments in general, and the mapping environment in particular
2	List any other software applications needed to support mapping
3	Install the software
4	Set up the mapping environment
5	Enter the data for baseline mapping purposes

application-level parameters and the setup in the model database may not represent the organization's business.

The rule is to quickly configure the base setup data as gathered from the business requirement definition phase and to identify the gaps to suggest possible workarounds to them. If workarounds are not possible, the justification has to be impressed upon the client, and the customization decisions have to be taken up jointly by the client, the implementation partner, and software owner. Do not try to perform the customizations in the mapping environment.

Each coded question on the "current business baseline questions for all modules" is categorized by process, performance, setup, or metrics. You can extract all setup responses and use this information as the baseline setup for mapping (Table 3.6).

The goal is to prepare the environment with the basic setups that allow the project team to begin the mapping and the client to understand and appreciate the mapping results.

For multisite replication, after one site has configured the mapping environment, capture or export the configuration of the application database and import the data to other site environments.

Map Business Requirement, Business data, and Identify Gaps

Mapping a business process means:

- Proving designs through demonstration
- Identifying gaps in the application
- Proposing feasible bridges to gaps

The following list includes some broader connotations of the term mapping:

- The basis for establishing application fit to business requirements, identifying gaps, and proposing alternatives
- The formal linkage of future process models to application features

3.3 Detail Activities

Table 3.7 Mapping process

Process	Core process analysis
Ref No.	Task
1	Train all assigned team members in the use of the methods and tools for mapping
2	Check out the prerequisite deliverables and become familiar with business requirement scenarios for the target process in need of mapping
3	Conduct mapping sessions to assess detailed application fit and create or revise alternative to business requirements. Map future business requirements to application features, programs, reports, and other standard modules
4	Perform online prototyping and deliver a prototype demonstration
5	Perform process research; look for and document alternatives
6	Identify current versus proposed process steps and assess the feasibility of proposed alternatives
7	Document alternatives. Record possible alternatives for application gaps
8	Document major operating and policy decisions

In this regard, mapping describes the evolution of process design for a business process. The business requirement definition will continue to evolve and be supplemented and improved throughout all mapping tasks (Table 3.7).

The formal mapping task, however, is very specific in that it documents key business requirements and proposed alternatives in much more detail.

When mapping, keep the following steps in mind:

- Address critical business processes identified by the organization before seeking resolution to minor issues that crop up in business process designs and maps.
- Use standard system features, functions, and reports whenever possible.
- Use extensibility features, such as descriptive flex field.

New business requirements could emerge during a mapping session. Verify that these new requirements fit within the scope of the project before adding them to the business requirement list. Set aside time to finalize these requirements.

Make sure to identify differences between true business requirements and a wish list.

It is important to involve business line managers in the mapping development process. This provides an excellent opportunity for leaders to gain practical experience configuring and testing the applications. As soon as possible, encourage the business line manager to take responsibility for driving mapping sessions, thus allowing the application specialist to facilitate and provide guidance.

Consider these prototyping tips:

- Quickly devise and show an essential alternative (not necessarily complete the first time).
- Focus on core processes and characteristics that drive business objectives.
- Create just the major components of a working model.
- Do not worry about cosmetics.

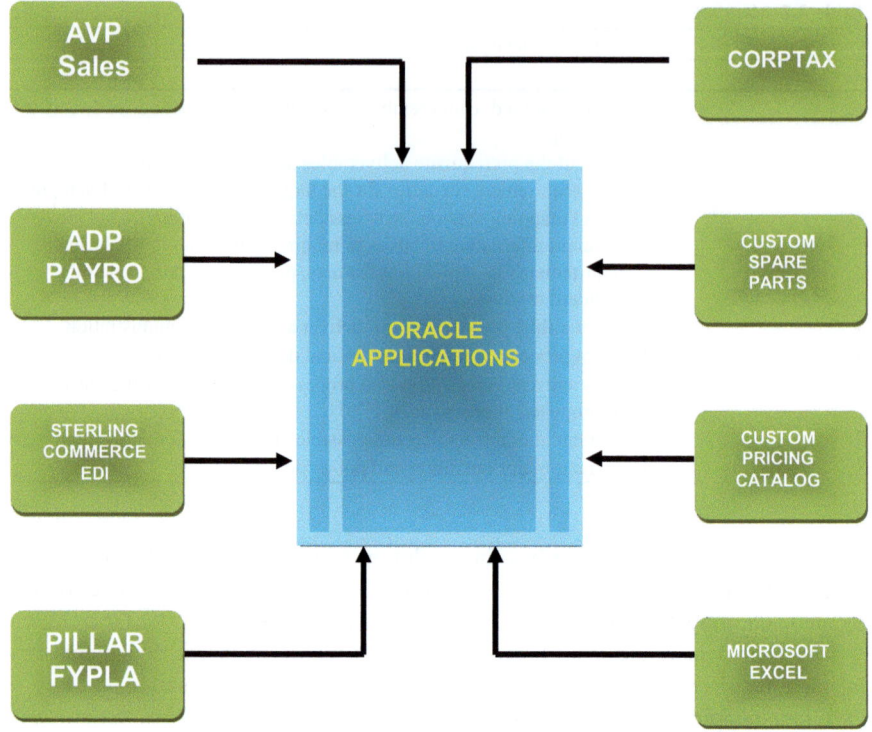

Fig. 3.3 Integrated fit analysis

Business Data Mapping

- Identify the legacy source file and field/data elements that are being converted and record this information.
- Identify the target application, business object name, and attributes that the legacy files and fields map to and record this information.
- Record the field that the legacy system stores but the application does not, and record the attribute that the application stores but the legacy system does not

The integration fit analysis describes how to identify and use integration points.

Figure 3.3 is an example of integration points in a business system.

The primary purpose is to map "distributed data external integration" points between standard applications and other third-party applications. Once you agree on an alternative for each integration point gap, this task will result in a complete listing of all new integration points for which you must design and build interfaces.

The Detailed phases of Interface Methodology are described within the Technical Implementation Methodology

3.3 Detail Activities

Table 3.8 Reporting

Process Ref No.	Core process analysis Task
1	Review the reporting strategy to understand the capabilities of placing reporting systems and constraints on designs
2	Decide on an approach for mapping report requirements and assign responsibilities
3	Map report requirements to standard application reports
4	Analyze reports for reduction
5	Prioritize custom reports

Conduct Reporting Fit Analysis

You may be able to employ the use of special reporting systems to reduce the number of reports you need to design and build. Examples of such systems are:

- Business intelligence systems
- Data warehouses (operational or decision support)
- Online analytical processing (OLAP) systems
- Ad hoc query systems

If the architecture work completed so far during the project has already identified the need for such systems, work with the system architect to understand how you may make use of these systems to satisfy reporting needs (Table 3.8).

The following are some of the typical report mapping issues:

Flex Fields Data captured in a flex field will not be part of a standard report; therefore, any report requirement using flex field data will become a *custom extension*. Sometimes, you do not know whether data will be stored in a flex field or another application field used by a standard report. In these cases, mark the report as a match with a note to modify the report, if the data are stored in a flex field.

Lack of Training Users are often trained just before the implementation is complete. Unfortunately, mapping occurs much earlier in the project. If users are going to do their own mapping, they will need the following:

- Access to a prototype environment
- Training on future processes
- Training on how to run reports

Some form of reduction process must take place when there are more custom reports identified for development than:

- Are necessary to run the business
- Can be completed in the allocated time
- Are expected, potentially placing reporting development over budget

The following are suggestions for reducing the number of report requirements:

- Eliminate reports with duplicate file names. Do not delete these requirements from the list, since they represent a valid user requirement that is necessary when preparing status documents for the user, user's department, or management.
- Analyze based on function. Resort the master report tracking list (current business baseline) by function. If several reports relate to the same function, you may be able to combine the requirements from one report into another.

Warning Track consolidations carefully, especially if they cross departmental boundaries. While initially all parties may agree that the consolidation looks good, changes requested by one group may not be appropriate for others. When this happens, you may need to create another new report and track it separately.

Gap Resolutions and Create Future Process Model

There are many types of alternatives to application gaps, ranging from large subsystems to localized modifications (configuration, setup, flex field, alert, report, and form), to simple workarounds.

The revised process reflects the approach dictated by the workaround and downstream users, and reviewers of the process are able to use the reasons and support information.

Consider these tips when mapping and creating gap resolutions:

- Design alternatives for the desired state of the business, rather than directly mapping to current needs.
- Always implement workarounds before designing and building a custom extension.
- When multiple alternatives are available, choose the alternative that supports organization goals or broad business areas, rather than satisfying the needs of a single department or user.

In order to obtain rapid implementation consider these mapping tips:

- Get quick closure on gaps.
- Push hard on perceived gaps—up to 80% of the initial gaps identified may actually be found to be unnecessary.
- Ask the question, "What does the application do?" in order to keep the mapping session moving.
- Adjust business processes to fit ERP application functionality.
- Pay special attention to prioritizing gaps in order to manage scope and budgeted resources properly.

Consider the decision flow diagram before arriving at any gap resolution. Each gap needs to be analyzed in light of the decision flow diagram followed by extensive discussion with the key process owners before arriving at a consensus and a collective buy-in of the business, systems, implementation partners, and supplemented by the software owner for a feasible bridge to the gap.

3.3 Detail Activities

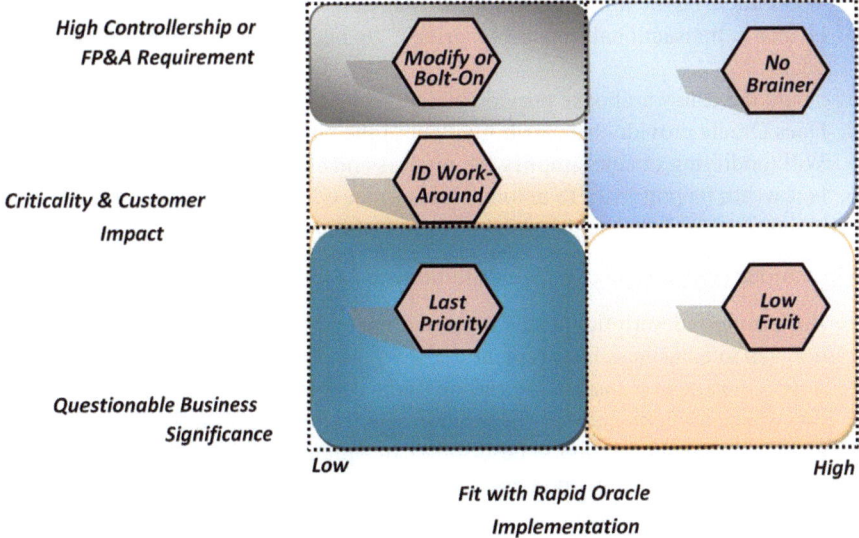

Fig. 3.4 Business flow diagram: core process analysis

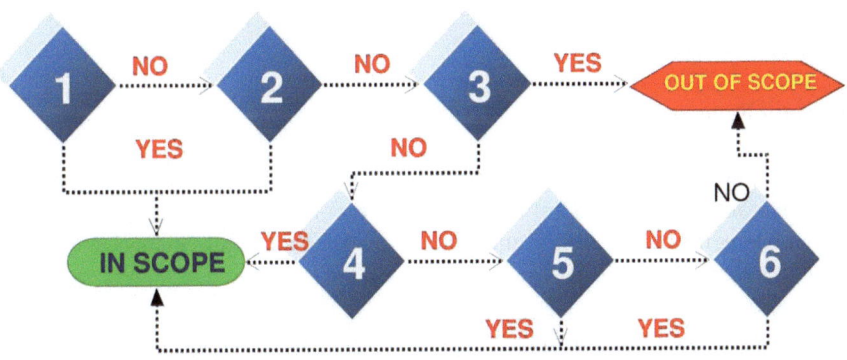

Fig. 3.5 Scope management second pass

Scope Management: First Pass (Fig. 3.4)

Scope Management: Second Pass (Fig. 3.5)

1. Will Oracle be the initial data input so that an interface will not be required?
2. Is this a transactional capability critical to operations, finance, or customer quality?
3. Could the data warehouse provide better "real-time" access to this information?
4. Does Oracle provide this capability "out of the box?"
5. Will modifying Oracle simplify the process and eliminate the parallel path?
6. Is it worth paying extra to automate this process? Is it efficient and clearly rule based?

The various components of a gap resolution are as follows:

- *Workaround:* description of the proposed method for getting around an application gap to a business requirement
- *Application enhancement:* description of the custom modification to the application whose implementation will result in satisfaction of the business requirement
- *Reengineering opportunity:* description of simplification, elimination, or enhancements of the target process
- *Solution/design document reference:* if available, a document number for high-level or detailed design planned to satisfy the requirement
- *Mechanism:* resources that influence the process; people, tools, or machines affected by the BRM proposal
- *Interfaces:* description of system interface requirements necessary to satisfy the requirement
- *Solution technique:* description and type of application feature that will satisfy the requirement (configuration, setup, flex field, alert, report, form, and so on)

Custom Reports Prioritization

As you map reports to standard functionality, custom requirements may develop. Anything marked as a *build* or *modify* is a custom requirement. Sort the master report tracking list by *assessment,* and make sure all custom requirements have a priority (Table 3.9). Print this list and distribute it to the team and users, and request that they make any necessary changes to the priority. This will be an ongoing function.

Priority is the basis for the drive for the entire development process and thus needs careful management. Users should always sign off on the assigned priority to avoid conflicts at later stages.

Confirm Integrated Business Solutions

The confirmed business solutions (Table 3.10) present the business requirement scenarios (modified current business baseline) and the mapped business requirements (future business model) to management for approval.

You may request approval for each process area as it is mapped, or you can get approval for an entire functional area (for example, manufacturing, finance, distribution, and so on).

3.6 Critical Success Factors

Table 3.9 Report prioritization

Process Ref No.	Core process analysis Task
1	Review the gaps in light of the business requirement and propose solution alternatives
2	Create an impact analysis of each proposed solution on the risk continuum
3	Create an impact analysis of each proposed solution based on incremental change in schedule and cost
4	Review and analyze the impact analysis information with the client management/client decision-making body
5	Analyze each gap through the scope management models as given above
6	Freeze a feasible consensus solution for each gap
7	Review the financial and operating structure, current business baseline including the master report tracking list in light of gap resolution and revise the documents if necessary
8	Create the future process model and review it with the client for acceptance
9	Prioritize the custom reports and custom components

Table 3.10 Integrated business solutions

Process Ref No.	Core process analysis Task
1	Review prototype and mapping documents
2	Revise business alternatives for agreed-upon changes
3	Prepare an acceptance certificate for integrated alternatives
4	Secure acceptance of the confirmed business solutions

If alternatives are not accepted, the reasons for nonacceptance should drive a round of revisions to the affected mapped business requirements deliverable. Another cycle of process modeling, process design, and mapping will be necessary, but unless the future business model and gap resolutions are frozen and accepted by the client, do not proceed to the next phase of implementation.

3.4 Deliverables (Table 3.11)

3.5 Decision Matrix/Checklist (Table 3.12)

3.6 Critical Success Factors

- Dedicated resources for conducting analysis
- Project team's understanding of application functionality and the leading industry practices

Table 3.11 Deliverable, template ID, and responsibility

Sl No.	Name of deliverable	Template ID	Responsibility	
			Major	Assisted
1	Current financial and operating structure		Consulting organization	Client
2	Current business baseline		Consulting organization	Client
3	Business volumes and metrics		Consulting organization	Client
4	Audit and control requirements		Consulting organization	Client
5	Business availability requirement		Consulting organization	Client
6	Key process owner-training plan		Consulting organization	Client
7	Training course attendance record		Consulting organization	Client
8	Business requirement mapping document		Consulting organization	Client
9	Impact analysis V0.0		Consulting organization	Client
10	Current business reports		Client	
11	Existing reference material related to gain an understanding of the existing practices, processes, and systems that support the organization		Client	
12	Existing reference material for audit and control policy's and procedures within the company		Client	
13	Preliminary business requirement scenarios existing in the company like the sales cycle or the procurement cycle, etc.		Client	
14	Current business contingency plan and procedures		Client	

- Active involvement and support of management
- Active involvement and support of knowledgeable business area specialists
- Full access to information about relevant business areas, their processes, data generation, and use
- Thorough review, feedback, and acceptance of the deliverables within the stipulated time frame by the business
- Definition of clear and realistic business expectations and organizational performance measures from the training sessions
- Visible support and participation of key leaders and sponsors throughout the impacted business units
- Mechanism to listen and respond to top concerns about the new systems
- Early establishment of ongoing communication and feedback/evaluation mechanisms that fit the organizational environment
- Management activities that help employees understand their new performance objectives and expectations, as well as the importance of their job to the change effort

3.6 Critical Success Factors

Table 3.12 Major activities, checkpoints, and weightage

Major activities	Checkpoints	Weightage
Business requirement definition	Are all relevant data and information asked from client available and complete to the project team?	
	Are all the relevant deliverables prepared?	
	Have the deliverables to the client been submitted for review within time?	
	Was the review feedback from client done within the stipulated time frame as communicated?	
	Was the incorporation of review feedback done?	
	Has the final deliverable been prepared?	
	Was the submission of final deliverable done with clear communication of the acceptance time frame?	
	Is the client acceptance of the deliverables complete?	
Key process owners training	Is the attendance according to the plan?	
	Was the training held according to the schedule?	
	Were all training materials distributed to all the respective participants?	
	Were client feedback and response collected?	
	Has necessary plan of action thought of according to the client feedback?	
Business requirement mapping and gap analysis/resolution	Is the hardware/software in place?	
	Is the business process-mapping document prepared?	
	Did the client review the deliverable?	
	Was the closure and consensus on gap resolution from client reached?	
	Review feedback from client is within the stipulated time frame as communicated	
	Was the review feedback incorporated?	
	Was the final deliverable prepared after freezing the future process model according to the business requirement-mapping document?	
	Is the final deliverable submitted with clear communication of the acceptance time frame?	
	Is the client acceptance and sign-off of the deliverables complete?	

- Consistency of team composition across process design, mapping, narrative writing, and approval activities
- Consistency of team approach across business areas or process groups
- Project team's understanding of primary and related applications and industry practices

- Active involvement and support of management and team access to business area experts
- Clear and concise visualization regarding how information will flow across and be owned by organizations, functions, and applications
- Effective management of issues, including timely resolution
- Proper sign-off throughout all stages
- Flexibility of the business to accept process change/work around instead of hard customizations to close the gaps

Chapter 4
Conference Room Pilot

4.1 Objective

The purpose of the Conference Room Pilot (CRP) is to install the CRP instance, configure the core application, and perform business transactions with the client's key process owners to establish that the basic client's business process framework is properly configured within the core enterprise resource planning (ERP) application installation at the client site (Table 4.1).

The CRP constitutes these activities:

Instance strategy

- Prepare instance strategy
- Install the CRP instance

Define application setup

- Collect global application setup data
- Prepare global application setup document
- Configure core application as per the setup data

CRP and testing

- Prepare testing strategy
- Perform unit testing
- Perform business transactions with the client on the application instance
- Identify any further gaps
- Conduct gap analysis and resolution
- Prepare for integration testing

CRP refers to the technique of setting up a conference room where the transaction tester's workstations are arranged in a particular order (usually by logical processes) and the test scripts are passed down the line from one tester to the next according to the natural flow of the business process. A CRP test usually involves performing a system test to check the validity of application setups, and the integration of business system flows within the target applications system (Fig. 4.1).

Table 4.1 Implementation schedule: Conference Room Pilot

Activities	TimeLine											
	Month1				Month2				Month 3			
	1	2	3	4	5	6	7	8	9	10	11	12
Prepare Instance Strategy	■											
Collect Application set up data		■	■									
Prepare application set up document			■	■								
Prepare unit test scripts				■	■							
Install ERP Application CRP instance				■	■							
Configure the Core application					■	■						
Unit test the application modules						■	■					
Perform the CRP with business process owners							■	■				
Identify further gaps, analyse and resolve								■	■			
Prepare for Integration Test									■	■		

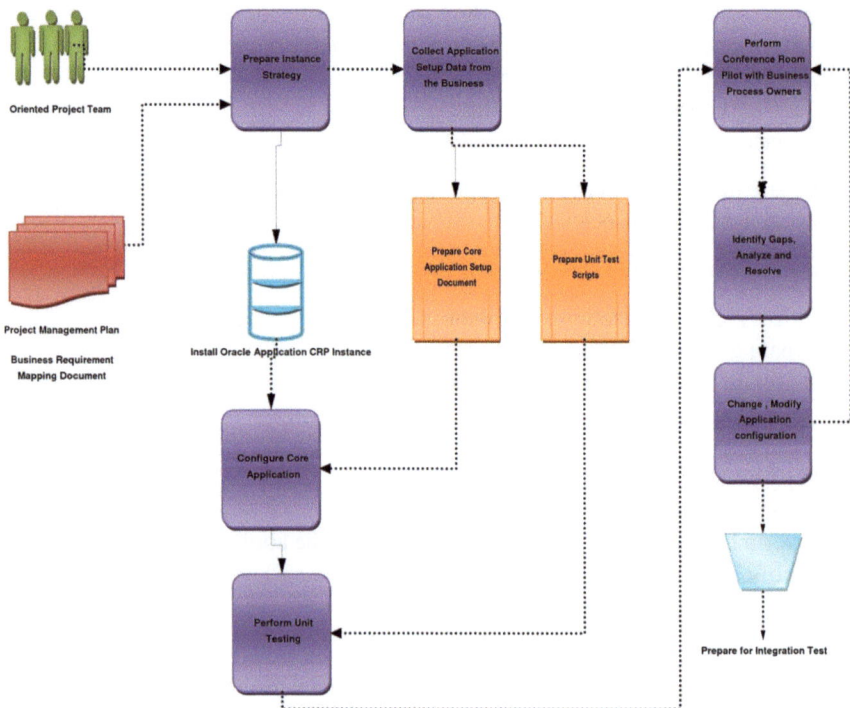

Fig. 4.1 Business flow diagram: Conference Room Pilot

4.3 Prerequisite 73

The purpose of the CRP is to test the operation and integration of the business system processes. Successful CRP should verify the following:

- The system successfully supports business operations.
- All results are predictable and repeatable.
- Variations of the test script produce expected results.
- Data are verifiable through alternative means across the business system.
- Relevant key performance indicators can be validated.
- Support procedures intended for production are accurate.
- Modifications to procedures, policy, and system are fully implemented.

4.2 Business Flow Diagram: Conference Room Pilot

4.3 Prerequisite

Prepare Instance Strategy:

- Experienced application and technical architecture practitioners
- Balanced input of business and technical requirements with business requirements driving the architecture
- Data conversion requirements
- Testing requirements and testing strategy
- Existing system architecture and policy documents

Define Application Setups:

- Process and mapping summary
- Gap resolution document
- Application architecture requirements
- Application setup templates circulated to the client
- Oriented project team
- Setup checklist availability as per the Application Implementation Wizard
- Frozen multi-org structure and SOB architecture accepted by client

CRP and Unit Testing:

- Global Setup documents prepared and reviewed
- CRP instance is ready for configuration
- Unit test scripts are prepared
- Early notification of managers that their staff (key users) will be involved in testing
- Test data available to be entered into the CRP instance

4.4 Detailed Activities

4.4.1 *Prepare Instance Strategy*

The objective of an instance strategy is to design information systems architecture to realize the business vision. A coherent and well-designed information systems architecture is critical for the success of any implementation project. To arrive at an architecture for the newly implemented systems, consider the following complex issues:

- The best deployment strategy for the applications across one or more data centres, business organizations, and business functions
- The high-level configuration of the applications to support the financial, administration, manufacturing, customer management, supply chain, selling chain, and distribution business units
- The interface points between the applications or remote sites, including specifications, data flows, and mechanisms
- The deployment and capacity planning of the hardware and network infrastructure that will support the applications processing
- The management tools and procedures that will enable the system to continue to operate as intended
- Backup strategy during system configuration and development

The tasks related to preparing the above document are as follows:

- Identify constraints and assumptions that are associated with the process
- Review risks to the process activities and objectives
- Define the environment requirements for the testing, conversion, learning, mapping, development, performance testing, and architecture activities of the project

4.4.2 *Prepare Backup Strategy*

A proper backup strategy has to be prepared for keeping the backup of instances during the development, CRP, and application deployment phase keeping in view the facilities and resources available for supporting the backup strategy.

4.4.3 *Define Application Instance Management Procedures*

One of the difficulties in discussing system management is that the area covers a diverse variety of system events and procedures. Although it is impossible to predict every conceivable event or situation in a complex system, the goal of the work performed in this task is to predict as many of them as possible. A well-designed system management strategy can pre-empt a great majority of system outages and prevent the necessity of designing procedures as situations unfold in development and transition activities.

The various areas, which need to be kept in view for defining a proper system management procedure, are as follows:

- Database management
- Security and accounts management
- Scheduling management
- Hardware and network management
- Software management
- Capacity planning and performance management

The administrator responsible for defining the procedures and tools for an area should consider the following areas and identify those that are relevant:

- Proactive and reactive monitoring
- Normal management and maintenance procedures
- System failure analysis and recovery
- Long-term resource planning
- Tools and utilities

4.5 Install the CRP Instance

4.5.1 Define Application Setup

Collect Core Application Setup Data and Prepare the Application Setup Document

The activities to collect the setup data for the core application configuration are as follows:

- Review business mapping decisions and documents
- Define the application setups intended for production and prepare the application setup template
- Fill up the application setup template prepared above with the setup-related data already available from the business requirement mapping documentation
- Make a list of the incomplete setup steps due to the want of setup data from the users
- Identify the specific users / key process owners who will be capable of providing the above data
- Discuss with the identified users / process owners and collect the relevant setup data
- Complete the setup templates and complete the *Global Application Setup Document*
- Review the setup documents with a view to cross module setup parameters

The application setup documents record the parameters, user-defined codes, and other setups for each application. The main objective is to consolidate the configuration of all applications for centralized maintenance. With the number of separate application databases on the organization's systems, it becomes a challenge to make sure that the configurations represent the latest mapping decisions. Only key project team members, each with specific responsibilities, should initially define the system application setups.

Critical Setup Parameters

Application parameters do not all carry the same importance to the business. Some are more critical in determining how the system will be operated. For instance, within the standard applications, the following parameters control significant processing features and functions:

- Set of books—an accounting ledger with a particular chart of accounts, functional currency, and accounting calendar
- Balancing entity—a division or other business unit for which you prepare a balance sheet
- Inventory organization—a plant, warehouse, or other type of business units designed to provide control and transaction functionality within one or more applications modules

In ERP applications, you have choices for the number of sets of books, inventory organizations, and installations of each application. Multi-organization functionality adds two further critical functional concepts:

- Legal entity
- Operating unit

This area also covers issues in the use of key application features that would affect application deployment. These decisions are dependent on the specific features of the application release.

Significant risk of implementation failure exists if the multi-organization feature is not adequately configured for the organization. Careful review of the documentation, available white papers, and the engagement of a qualified multi-organization architecture professional can reduce the risk of failure.

Collect Setup Data for Security Profile of the Organization and Complete the System Administration Setup Document

- Define the application setups intended for production and prepare the application setup template from the system-administrator perspective
- Identify user roles across all business functions and organizations

4.5 Install the CRP Instance

- Identify security requirements for each user role
- Map user roles onto application security structures
- Define application module access for each system user role

The security profiles' primary objective is to develop a security structure that naturally supports business processes. The primary technique is to map business process steps and their agents (owners) with the application-provided user responsibilities and to make adjustments to the responsibilities as required.

It is important to achieve a good menu structure so that application access naturally supports the flow of information in the workplace. This will also make learning the application easier.

Configure the Core Application Modules in the CRP Instance and Prepare for CRP

Review the application setup documents and put them in a version control mode through some tools like visual source safe (VSS), etc.

Prepare a checklist of setup steps (*Setup Checklist*) as per their order of precedence like

- Common application setup steps
- Common financial setup steps
- Common distribution setup steps
- Common manufacturing setup steps
- Individual module setup steps

Identify and communicate responsibility of each team member for configuring the relevant setup steps in the above checklist.

Monitor the configuration activities and the cross module configuration in the order of precedence.

Complete the configuration of the core application modules and system administration module as necessary as per the setup checklist.

Review and confirm configuration and update impact of changes in the application setup documents.

Strictly version control the global setup documents.

Application Implementation Wizard

The Application Implementation Wizard (AIW) is ERP's workflow-based setup tool that guides you through the setup steps in the appropriate sequence. The project manager can consider the use of the AIW tool from ERP for completing the application configuration, if necessary.

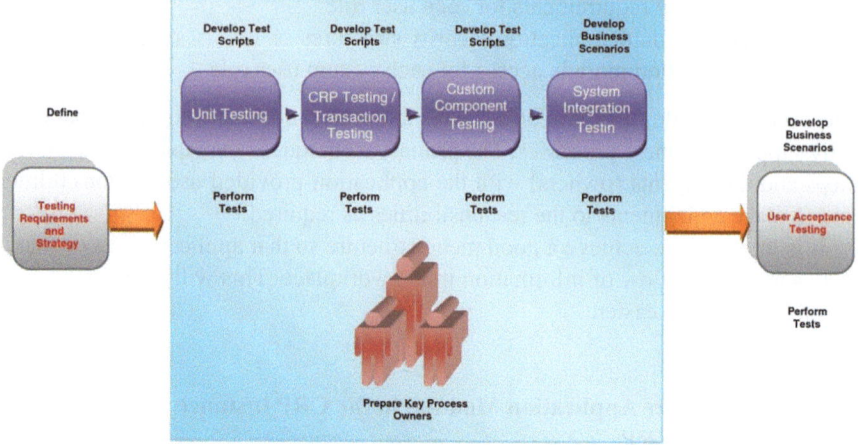

Fig. 4.2 Logical progression of the Business System Testing

4.5.2 Unit Testing and CRP

Prepare Testing Strategy

The objective of Business System Testing is to test the quality of all the elements of the application system. Business System Testing emphasizes a common planning approach for all types of testing and advocates the reuse of deliverable components to test successively larger aspects of the application system.

Figure 4.2 shows the logical progression of the Business System Testing tasks:

Business System Testing starts at the smallest element—the unit test—and expands to include the CRP / transaction test, the custom component test, and the systems integration test. For those environments where there are no custom extensions and no interfaces to legacy or third-party systems, there is no need to perform custom component test and systems integration testing.

In the present phase, the testing process starts with the unit testing phase.

The detail activities are as follows:

- Review the Project Management Plan and verify the testing description in the project-level scope, objectives, and approach contained within the Project Management Plan
- Identify the scope of the testing, types and iterations of tests, and relationships to other projects and systems
- Identify the high-level approach for carrying out testing activities
- Identify the testing requirements at the detailed level
- Define the resource requirements for the testing activities of the project
- Develop the testing strategy for the project

4.5 Install the CRP Instance   79

The testing strategy for the project is a global document, which includes the approach and plan and techniques for all types of testing as per the logical testing flow diagram as shown above. Unit testing and CRP testing strategy is a part of the overall document.

In the testing requirements and strategy phase, you establish or provide the following:

- A list of testing requirements
- An overview of the strategy, including relevant background, the testing approach, critical success factors for successful testing, and the risks associated with not performing adequate testing
- An understanding of the type and purpose of each testing task
- An understanding of the deliverables for each testing task
- An overview of the testing tools
- An overview of how problems will be managed
- Detailed acceptance criteria for testing

The objective of this task is to prepare a working strategy document that team members can reference during the tasks in Business System Testing. New testing team members can quickly become familiar with the process by reviewing this document; it can also resolve any questions about general information listed in the Project Management Plan.

Prepare Testing Environment/Test Instance

The testing requirements and strategy provides the details describing the environments required supporting each testing task. Configure the test environments based on the latest application setups from the Global Application Setup Documents. Before you begin testing in each environment, you should:

- Set up the applications
- Decide on the real business data to use
- Plan the volume of data to be entered during testing
- Verify that all application parameters have been set up to enable transactions
- Provide access to definition and setup screens to appropriate users
- Record all changes and updates to the test environment in the Test Environment Setup Log template to help you maintain the integrity of the objects that have been installed or updated

You may need to create multiple test environments to accommodate the various types of Business System Testing. Instances should be set up for unit test CRP / transaction test—this should be installed after the updated configurations are done as a result of unit testing.

The CRP test instance should include the following components:

- Configured server with application
- Configured database

- Loaded manual data, mainly various master data loaded during unit testing of modules
- Loaded help text (optional)

Perform Unit Testing

The unit test is the narrowest scope of testing you will perform. Each unit test script exercises a single module within the configured application in the previous step. When performed thoroughly, unit testing is one of the biggest contributors to a stable application system and will significantly reduce all downstream testing efforts.

Unit testing is a repetitive task; testers execute each unit test numerous times using different combinations of test data as specified in the data profile of the unit test scripts.

Figure 4.3 depicts the flow of unit testing process:

Organize your tests to evaluate normal usage cases first and then test exceptional or out-of-range cases and boundary conditions. A boundary condition is a combination of input parameters that represent the largest or smallest values permitted. It is much easier to diagnose problems during unit testing rather than later when many different programs are interacting.

The activities performed under this step are as follows:

- Review the business requirement documents for identifying possible test scenarios
- Develop unit test specifications
- Develop the data profile for each test case
- Review the unit test script
- Using the unit test scripts, identify the functional items and the cosmetic items to be tested
- Perform the unit test and document the unit test results in the actual results column
- Fix errors in the application extension
- Repeat the unit tests (if needed)
- Update the setup documents as per the unit testing feedback and response and update the version of the global setup documents

Conduct CRP and Prepare for Integration Testing

The detailed activities in CRP are as follows:

- Generate the business transactions specifications document, which will be used during the CRP testing including the desired results for each transaction flow
- Review the responsibilities for the CRP test with the team
- Validate the CRP test environment
- Prepare the key process owners for the CRP testing process
- Execute the CRP test script
- Document the detailed test results
- Review the results with the process owner

4.5 Install the CRP Instance

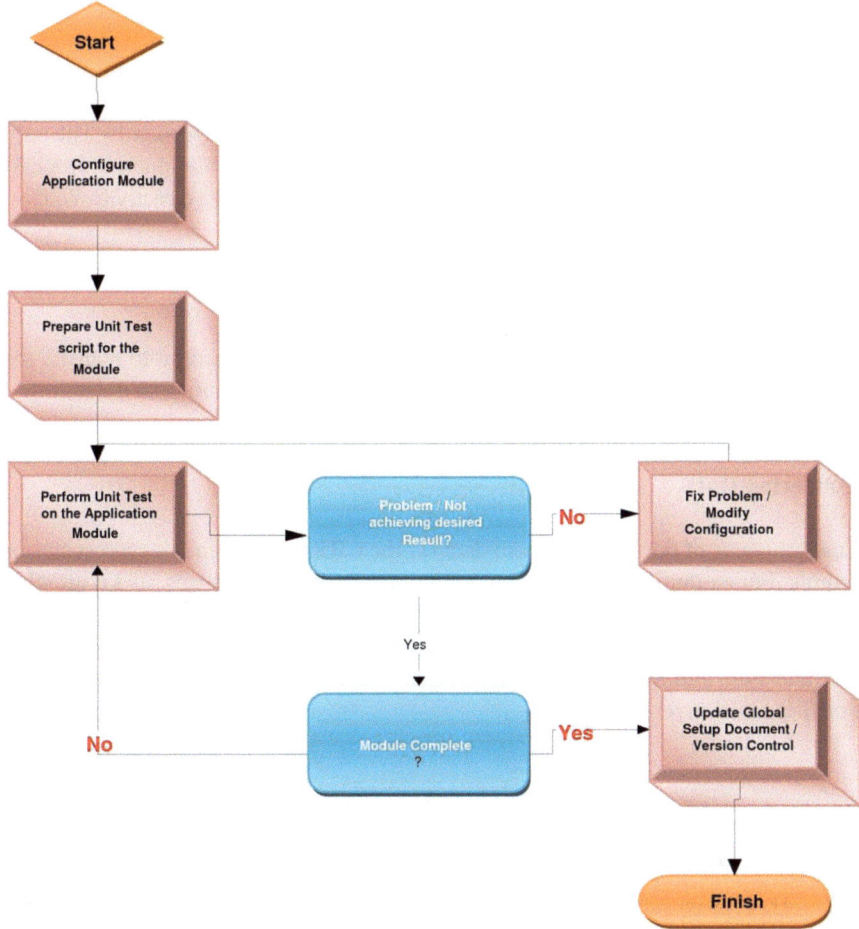

Fig. 4.3 The flow of unit testing process

- Identify new gaps, if any
- Determine required actions for gap resolution
- Make recommendations for applications changes or corrections based on system testing.
- Modify the application configuration
- Update the Global Application Setup documents and update its version

During CRP testing, users develop confidence in the programs and the overall usability of the target application system. It is important to test business processes and transactions performed by users in day-to-day circumstances, thereby simulating business operations, not just systems.

The repetitive nature of this task allows the project team to test each business flow and incrementally improve the system until they reach the desired level of success.

Parallel Process Flows

Where processes overlap or are independent, it is possible to perform process testing in parallel with coordination. For example, processing purchase orders and building items in work in process (WIP) can be performed independently. The only coordination is at the points when the processes intersect, such as outside processing of WIP operations or receiving material on a purchase order directly to WIP.

Test Condition Recreation

When attempting to validate that a defect has been fixed, be careful to recreate the same data conditions and sequence of events that exposed the original defect.

Converted Data

Convert a part of the legacy system data for the CRP test, which is a representative sample. For example, if you are converting general ledger balances, make sure that you include converted journals from each cost center, business unit, division, and so on. If you are converting vendors, make sure that the data include vendors with multiple sites and multiple contacts, so that you can test full vendor-related functionality and conversion.

4.6 Deliverables (Table 4.2)

4.7 Decision Matrix/Checklist (Table 4.3)

4.8 Critical Success Factors

The critical success factors for preparation of the CRP are as follows:
- Early initiation of an architecture process within an implementation project
- Realistic consideration of the capabilities and the limitations of the technology to be used
- Experienced application and technical architecture practitioners
- Balanced input of business and technical requirements with business requirements driving the architecture

4.8 Critical Success Factors

Table 4.2 Name of the deliverable, template ID, and responsibility

Sl no.	Name of the deliverable	Template ID	Responsibility	
			Major	Assisted
1	Global Application Setup Document Version X.0		Consulting organization	Client
2	Setup checklist		Consulting organization	Client
3	Unit test specifications for module		Consulting organization	Client
4	Business process scenario specifications		Consulting organization	Client
5	Setup data		Client	
6	Acceptance of agreed upon organization structure mapped to the application's multi-org environment		Client	
7	Acceptance of agreed upon SOB architecture		Client	
8	Provide/prepare the business process scenarios that are required to be tested during a CRP Test		Client	

Table 4.3 Major activities, checkpoints, and weightage

Major activities	Checkpoints	Weightage
Prepare instance strategy	Application deployment plan is available	–
	Backup strategy is available	
	Facilities and infrastructure support for backup strategy is available	
	Proper hardware is available for supporting the instance strategy	
	Instance strategy does not affect the project schedules and go-live date	
Define application setups	The impact of the multi-org structure understood and accepted by the client	–
	The impact of the SOB architecture understood and accepted by the client	
	All setups' data collected from the client and documented in the Global Setups documents	
	Global Setups documents reviewed and controlled	
	Cross-functional modules dependencies identified, understood, and properly documented by the project team	
	Configuration completed as per setup checklist on the CRP instance	
Unit testing and CRP	Testing strategy is documented and reviewed	–
	Unit test specifications completed and reviewed	
	All modules fully tested as per the unit test specifications	
	All actual test results documented in the unit test specifications	
	All bugs and errors are corrected	
	Business process scenarios are complete and accepted by key process owners as CRP testing baseline	
	CRP testing completed and ratified by the process owners	
	Global setup documents are updated and the version controlled	

- Interactions with other tasks and activities of the project that impact the final production application or technical architecture or have interim architecture requirements in support of developmental activities
- Proper updating and strict version control of the Global Setup documents
- Proper monitoring of the sequence of setup steps as per the setup checklist
- Perfect coordination among cross module functions related to relevant dependencies during configuration
- Correctness of the multi-org structure being configured in alignment to ultimate business goals
- Correctness of the SOB architecture being configured in alignment to ultimate business goals
- Reliable approach for identifying system requirements
- Adequate resources and time for test-script development and execution, and support for testing environments
- Adequate tools, including properly configured environments and well-trained users, to support test script development and execution
- The development of test scripts that are based on business process-driven requirements
- Early notification of managers that their staff (key users) will be involved in testing

Chapter 5
Customizations

5.1 Objective

Enterprise resource planning (ERP) application provides the users, the flexibility to customize the standard application functionality to suit the specific business needs, which is not supported by the standard application product. Developing additional reports as per business requirement of the organization also falls under the scope of customization.

During any implementation of ERP application, a general observation is that customers request for a sizable degree of customization. Before doing any customization, it is the responsibility of the application consultant to ensure the following:

For Process Level Customizations

1. Whether standard applications do not provide that functionality
2. Whether it is possible for customers to fit into standard functionality if that is a better practice
3. Whether customization is coming as future release in ERP

For requirements related to reports, the approach should be to identify the fields and to find if any customization is required to satisfy the additional data requirement in the report. Figure 5.1 shows the customization required fields.

5.2 Process Flow Diagram: Implementation of Customization

5.3 Implementation Schedule—Customization

Table 5.1 shows the implementation schedule for customization.

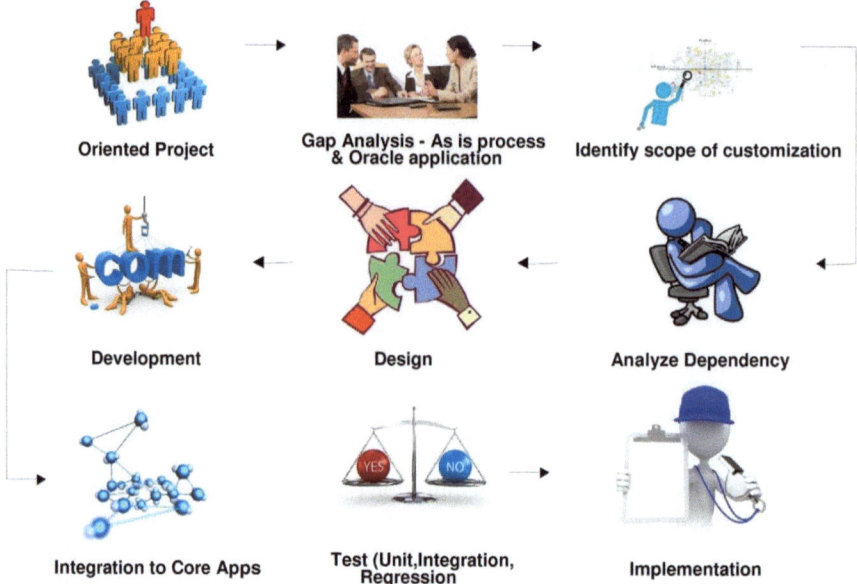

Fig. 5.1 Customization required fields

Table 5.1 Time line (Months)

Activities	TimeLine																
	Month1				Month2				Month 3				Month 4				
	1	2	3	4	5	6	7	8	9	10	11	12	13	14	15	16	17
Gap Analysis																	
Analyze Dependency																	
Design																	
Development																	
Unit Testing																	

5.4 Prerequisite

- TO BE process has been mapped with ERP application standard functionalities.
- Gap has been identified for all ERP application modules to be implemented.
- Need of customization has been decided in the steering committee meeting for GAP resolution.
- Customization analysis and customization-building team has been identified.

Table 5.2 Process for customization requirement

Process Ref No.	Customization Task
1	Identify process gaps
2	Finalize whether customization is required
3	Carry out feasibility study and estimate effort

5.5 Detail Activities

5.5.1 Gap Analysis

In this phase, a detailed study of the client's AS IS process is required, keeping the core ERP application functionality in the background.

We need to categorize the processes in each functional module to

- The ones which ERP supports as standard functionality
- The ones which are not there in ERP application

For the ones not supported by ERP, we need to identify the alternate solution available in ERP and discuss with client's functional owner of the process. The outcome of the discussion can be as follows:

- Client accepts ERP functionality as its TO BE process—no customization.
- Client cannot adapt to ERP-suggested process—customization required.

During discussion with the client, the following points should be stressed on:

- Customization requires extra cost of support, as ERP does not provide support to customization.
- Each time a patch is applied to an environment, there is a risk of losing customization. Therefore, rework effort is more.
- ERP suggests not to customize the process flow for transaction-related standard application. This is because of the complex nature of integration across different modules of ERP application.

Once we decide that a customization is required, a feasibility study is done and the effort is estimated. With the outcome, we move over to the analysis phase. Refer "Gap Resolutions and Create Future Process Model section of Core Process Analysis." Table 5.2 shows the process for customization requirement.

5.5.2 Analyze Dependency

We arrive at this phase with the outcome of feasibility study and effort estimate. This is the analysis phase of the customization where we identify the areas in core application which we need to touch (see Table 5.3). The major tasks of this phase are:

Table 5.3 Analyzing dependency

Process Ref No.	Customization Task
1	Prepare list of objects to be customized
2	Prepare process flow diagram for customization
3	Prepare analysis document
4	Prepare list of redundant standard functionality

Table 5.4 Approach during design

Process Ref No.	Customization Task
1	Prepare customization strategy document for all category of custom components
2	Prepare coding standards
3	Prepare design documents

- Determine detailed functional process flow diagram for the customization
- Identify and classify the objects that need customization (reports/forms/procedures/workflow/alert)
- Identify data flow link between customized object and standard applications objects
- Identify if there is any standard applications functionality that becomes redundant for the customization

5.5.3 Design

In this phase, we define the technology architecture of the customization (See Table 5.4). At the end of this phase, the deliverable is the design document.

The following approach can be followed during the design phase:

- For the object/functionality to be customized, check for closest match standard applications object/functionality which will be obsolete. This can be obtained from the analysis document
- Take the standard applications object as a baseline for customization
- Define the custom application which will own the customization

5.5.4 Development

This phase is instrumental in building the customization referring to the design document (See Table 5.5).

5.8 Critical Success Factors

Table 5.5 Development of the document

Process Ref No.	Customization Task
1	Build custom components as per design document
2	Prepare unit test plan and unit test case
3	Unit testing of custom component
4	Fix all the bugs identified at the time of unit testing
5	Update detailed design document

Table 5.6 Name of deliverable, template ID, and responsibility

Sl No.	Name of deliverable	Template ID	Responsibility	
			Major	Assisted
1	Functional requirement analysis document		Consulting organization	Client
2	Detailed design document		Consulting organization	Client
3	Unit test plan and unit test case		Consulting organization	Client
4	Source code of custom component		Consulting organization	Client

Table 5.7 Major activities, checkpoints, and weightage

Major activities	Checkpoints	Weightage
Gap analysis	Whether end-to-end business process has been covered for gap analysis	
	Whether the exception criteria have been taken into consideration	
Analyze dependency	All customizations identified	
	Feasibility study and estimation done	
Design	Functional requirement reflected into design	
	Technical review done for the design	
Development	Coding standard maintained	
	Code is reviewed	
	Unit testing done before formal delivery for client testing	

5.6 Deliverables

5.7 Decision Matrix/Checklist

5.8 Critical Success Factors

The critical success factors for implementation of ERP application customization are as follows:

Resource

- Gap analysis phase should have 90 % of the functional resource and 10 % techno-functional resource.

- For study of dependency phase, we should have 60% techno-functional resource and 40% functional resource.
- For design phase, we should ideally have 20% functional resource, 30% technical resource, and 50% techno-functional resource.
- Development phase should have 10% functional resource, 70% technical resource, and 20% techno-functional resource.

Complexity

- Should try to minimize customization of major transaction screen.
- Criticality of report customization is less as compared to process customization. This should be taken into consideration during estimation.
- For workflow customization, trained people are required on workflow development.

Analysis

- Correct and to-the-point end-to-end gap analysis should be the key to success of customization.

Chapter 6
Interface and Conversion

6.1 Objective

Although enterprise resource planning (ERP) application is meant for capture and to process the entire gamut of enterprise business, there are areas where a bridge between an existing application and legacy application is very much required. These are required in the following two conditions:

1. ERP application is replacing a legacy application that the business was using. Hence, the data need to be moved permanently into ERP applications. In this case, we talk about data migration/*conversion*.
2. ERP application will be running along with an existing legacy application. Hence, there needs to be a bidirectional data flow between the legacy system and ERP application. In this case, we talk about *interfaces*. Interfaces are *inbound* when the data are loaded into ERP applications and *outbound* when the data are extracted out of ERP applications.

In brief, conversion is usually meant for programs which are run once only, and interfaces are those which are run at a predefined frequency.

During the implementation of ERP applications, these interface and/or conversion programs serve as a bridge between the ERP application and the legacy system.

6.2 Business Flow Diagram

The data flow for inbound and outbound interfaces is depicted in Fig. 6.1.

The basic steps involved in inbound interface are:

1. Extracting data from the legacy system (Mainframe/AS400/legacy ERP database) to a flat file.
2. Loading data from the flat file to a staging area (separate database/schema/table) using custom SQL*Loader script or tools like DataMirror, MQSeries, etc.

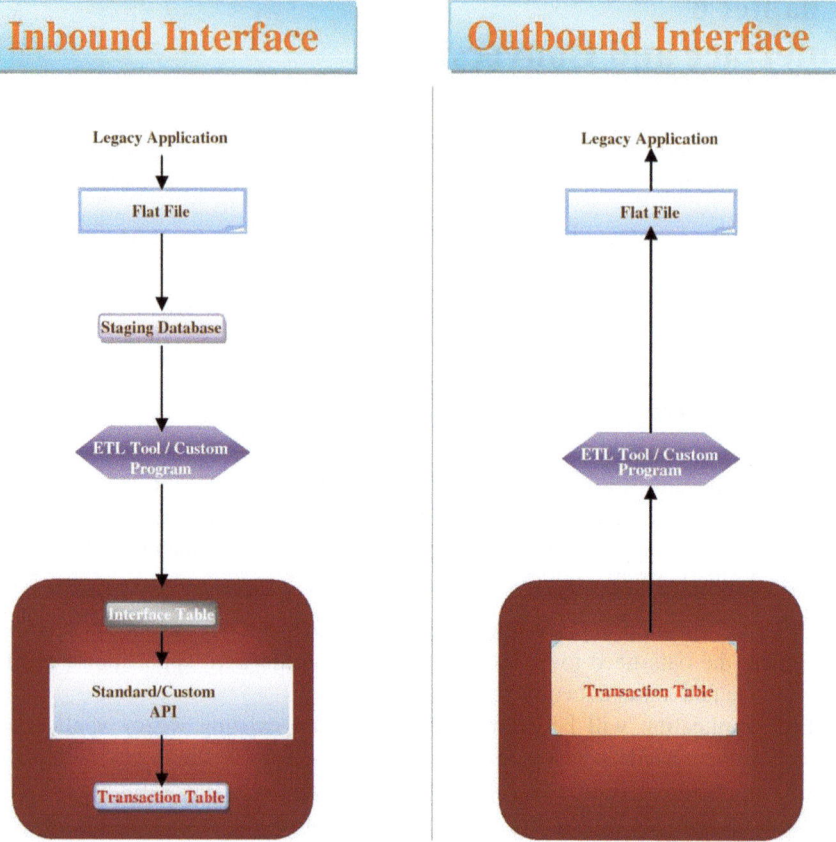

Fig. 6.1 Data flow for inbound and outbound interfaces

3. Validating and transforming data and loading them into ERP application interface table using custom program or tools like *Informatica,* Web method, etc.
4. Loading data into the transaction table using ERP standard API or custom-built API.

In case of outbound interface, usually the transaction data are directly extracted into a flat file, which is then passed on to the legacy system.

There are four major phases in an interface/conversion project. These are:

1. Requirement study/analysis
2. Design
3. Development and testing
4. Implementation.

The major activities performed in each phase are indicated in Table 6.1.

6.5 Detail Activities

Table 6.1 Activities and time line (months)

Activities	TimeLine												
	Month1				Month2				Month3				
	1	2	3	4	5	6	7	8	9	10	11	12	13
Identify Interfaces & Tools													
Prepare Interface Building Plan													
Interface Designing													
Interface development													
Interface Integration testing													

6.3 Implementation Schedule—Interface

6.4 Prerequisite

- TO BE process has been mapped with ERP application standard functionalities.
- GAP has been identified for all ERP application modules to be implemented.
- Need of interface program has been decided in the steering committee meeting for GAP resolution.
- Interface analysis and interface-building team has been identified.

6.5 Detail Activities

6.5.1 Identify Interfaces and Tools

The first step of requirement study is to identify the source and target area from which the data flow will take place. It is recommended that a high-level data flow diagram be drawn for each major area of data transfer at this stage. This would help to identify the potential gaps (if any) between the legacy system and the application. In this phase, the analysts should establish the following:

- A list of interfaces to be built.
- An understanding of the type and purpose of each interface.
- An idea about the deliverables in this phase.
- An overview of the scope and boundary for each interface.
- The tools that can be used.

The objective of this activity is to come up with a concrete boundary and scope of the project and identify the list of deliverables. Table 6.2 lists the tasks that need to be performed.

Table 6.2 Analysis of process and interface and conversion

Process Ref no.	Interface and conversion Task
1	Identify the legacy system with which the interface should be built and its hardware and software
2	Identify what are the different interfaces and establish their number and complexity
3	Prepare detailed data flow diagram for each interfaces indicating the direction (inbound/outbound), the frequency of each transfer and the expected volume per run
4	Comparative study between the various tools that can be used
5	Brainstorming session to find out the best possible tool and compare the pros and cons of utilizing it vis-à-vis custom written code
6	Have a final decision about the tool to be used or the custom programs to be written
7	Define the template of the design document
8	Establish what are the various open interface table involved and whether there is any need for custom written API

Table 6.3 Preparing interface-building plan

Process Ref no.	Interface and conversion Task
1	Define the resource plan for various phases of the project as well as their location (on site/offshore)
2	Define roles and responsibilities for various members of the team
3	Identify the development, test, and the production instance
4	Identify how version control of the code and document would be implemented
5	Have a sign-off from the client regarding scope, duration, effort, and elapsed time for the project

6.5.2 Prepare Interface-Building Plan

In this phase, the major task is to identify the resources of the development team and establish the point of contact for the legacy application (Table 6.3). This would ensure that the team in the development phase knows whom to contact to resolve any queries at that stage. It is expected that this phase should provide

- Detailed resource plan, offshore and on site, if necessary
- The effort and the elapsed time that would be required to complete the project
- The breakdown of the client's and the consulting organization's responsibilities
- The potential risks and contingency plans
- The target-quality metrics and service-level agreement (SLA) for the project.

6.5.3 Interface Designing

At this stage, the project is in the design phase and the team would be responsible to freeze out the structure of the source and the target data files and find out the

6.5 Detail Activities

Table 6.4 Designing the interface

Process	Interface and conversion
Ref no.	Task
1	Identify the fields in the source system that would be used for target data population
2	Design the data structure for the source and target data files
3	Define the business logic for the mapping between the source and target data files
4	Define any lookup tables that would be used to store the mapping to be used to compute the derived field
5	Specify the error handling procedure
6	Collect and store the reference data
7	Collect sample data files from the source system
8	In case of ETL
	Install the software
	Set up the environment so that the tool is ready for use in development
	In case of custom-written program
	Design custom schema with details of logical objects and physical storage
	Design validation routine for loading into ERP interface tables
9	Prepare detail design document for ETL tool usage and/or custom-written code
10	Take sign-off from the client

mapping between the fields of the legacy system and ERP applications. The design phase should also specify the error handling routines and may also specify a detailed pseudo-code of each of the program. Since, at this stage, it is decided whether any extract, transform, load (ETL) tool will be used and/or custom application code is to be written, the design document should be specified accordingly. In case where an ETL tool is used, this step would involve the installation of the software and necessary setting up of the environment. For cases where custom code is to be written, this step would define the schema that will hold all custom objects required for the purpose. The normal practice is to have a separate custom schema for each module that will hold all the codes, and applications will be granted necessary privilege on such objects. The various tasks that need to be performed are listed in Table 6.4.

6.5.4 Interface Development

At this stage, it is almost time for the development team to plunge into detailed coding. But before that, one should remember to define the naming standard and other coding standards for the entire project. This coding standard should be subject to peer review as well as external quality assurance check, as this defines the framework on which the entire coding would be done. Based on the design document prepared in the previous phase and on the pseudo-code written there, the coding and preparation of the unit test plan should be started. Each code should be subjected to a thorough peer review. During the review, the International Quality Accreditation (IQA) should check the code from the point of view of both business logic and the programming and naming standard. Also, during this phase, the required effort should be made to prepare the installation guide (Table 6.5).

Table 6.5 Development of the interface

Process Ref no.	Interface and conversion Task
1	Define and develop programming standard
2	Review and release the standards
3	Prepare unit test plan
4	Develop interface and/or custom API
5	Carry out unit testing of the program developed as per the test plan
6	Have a peer review of the code
7	Log defect and do a causal analysis for the same
8	Fix issues and release program for source code control
9	Prepare the installation guide

Table 6.6 Interface integration

Process Ref no.	Interface and conversion Task
1	Prepare integration test plan
2	Prepare test data to carry out end-to-end testing
3	Install the software using the installation guide
4	Carry out integration testing and log test results
5	Carry out volume testing with live data
6	Monitor performance during volume testing
7	Have a dry run on the sample data file collected earlier
8	Fix any bug that was detected at any stage
9	Update the installation guide, if necessary
10	Verify after uninstalling and reinstalling the software

6.5.5 Interface Integration Testing

The purpose of the integration testing is to do an end-to-end testing of the entire system. It differs from unit testing due to the fact that, here, the test plan should be developed driven by business functionality rather than by individual programs. This is the last chance for testing the code before it goes to the client, so one should consider volume testing as a part of this phase. Some mock dry run with sample business data collected in previous phase is extremely important. The tasks that need to be performed are listed in Table 6.6.

6.5.6 User Acceptance Testing

After the code is delivered to the client, it would be subjected to user acceptance testing. For this, a user acceptance plan must be prepared and the code should be tested to perform satisfactorily for all cases as in the plan. Any deviation should be logged and examined to find out whether this is due to a bug in the code or due to

6.5 Detail Activities

Table 6.7 Process for the user acceptance

Process Ref no.	Interface and conversion Task
1	Prepare user acceptance test plan
2	Prepare test data to carry out user acceptance testing
3	Install the software using the installation guide
4	Carry out acceptance testing and log test results
5	Carry out volume testing with live data
6	Monitor performance during volume testing
7	Fix any bug that was detected at any stage
8	Update the installation guide if necessary
9	Verify after uninstalling and re-installing the software
10	Update all deliverable document
11	Get signoff from client

Table 6.8 Process for identifying and freezing source data files

Process Ref no.	Interface and conversion Task
1	Prepare and review detailed data migration plan
2	Determine and freeze source file for historical transaction (e.g. GL YTD account balance)
3	Determine and freeze source file for master data (e.g. vendor, customer, item etc.)
4	Determine and freeze source file for open transaction data (e.g. open customer invoice, open vendor invoice, open purchase data etc.)

a change in business requirement. All changes should be controlled by consulting the organization to own change control procedure. The various tasks that need to be performed at this stage are listed in Table 6.7.

6.5.7 Identify and Freeze Source Data Files

Since the legacy system is still active, the client should agree upon the production cutoff date up to which point transaction data would be migrated. Once this is finalized, the source data files should be extracted and frozen. The various tasks involved are given in Table 6.8.

6.5.8 Load and Validate Source File for Data Migration

This is the most important step in the whole project as this activity encompasses almost the entirety of the data loading. One thing that needs to be kept in mind is that the master data should be loaded prior to the transaction data. Any limitations/

Table 6.9 Process for loading and validation

Process Ref no.	Interface and conversion Task
1	Identify dependencies and prepare the schedule of data loading
2	Load and validate historical transaction through conversion interface
3	Load and validate master data through master data interface
4	Load and validate transaction data through conversion interface

Table 6.10 Process for validating backlog transaction data

Process Ref no.	Interface and conversion Task
1	Freeze backlog transaction data between production cutoff date and go-live date
2	Load and validate backlog transaction data through ongoing interface

Table 6.11 Deliverables and responsibilities

Sl no.	Name of deliverable	Template ID	Responsibility	
			Major	Assisted
1	Interface analysis document		Consulting organization	Client
2	Interface design document		Consulting organization	Client
3	Installation and registration guide		Consulting organization	Client

dependencies of the API should be identified before the schedule of data loading is finalized upon. The various tasks that need to be performed are tabulated in Table 6.9.

6.5.9 *Load and Validate Backlog Transaction Data*

Since there is a finite gap between the production cutoff date and the go-live date, all data generated in the source system after the production cutoff date but before the go live date should be migrated. This is taken care of in this phase after which the interface runs at the frequencies identified before. The various tasks that need to be performed are tabulated in Table 6.10.

6.6 Deliverables

The deliverables at each phase of the project are mentioned in Table 6.11.

6.8 Critical Success Factors

Table 6.12 Process, checkpoints, and weightage

Process	Checkpoints	Weightage
Requirement study	Is the legacy system to be interface identified?	
	Is the tool to be user identified?	
	Have the number, frequency, and the complexity of interface been documented?	
	Have the resources been identified for analysis?	
	Has the owner of the legacy source data been identified?	
	Has it been agreed upon whether the extract from source system will be given to the consulting organization or is consulting organization expected to extract that as well?	
	Has the timeline for various phases and the effort been agreed upon with the client?	
Design	Have the data structures for the source and target data files been frozen?	
	Has the lookup table been defined?	
	Have the reference data sets been identified?	
	Is the error handling logic documented?	
	Has the business logic been understood and mapping between the source and target data sources been established?	
	Has the responsibility for data cleaning been defined?	
Development	Has the coding standard been released?	
	Is the development for all modules completed?	
	Have all bugs reported by peer review been closed and verified?	
	Is the installation guide ready?	
Testing	Are integration testing and user acceptance testing (UAT) complete with all errors tracked to closure?	
	Is the performance during volume testing satisfactory?	
	Have the clients given the user acceptance letter?	
Production Go Live	Are the production cutoff date and go-live date agreed upon with the client?	
	Have all codes and documentations been baselined and put under version control?	

6.7 Decision Matrix/Checklist

The checkpoints and their weightage are given in Table 6.12.

6.8 Critical Success Factors

The critical success factors for interface development project are as follows:

- A detailed project plan defining the effort, elapsed time, risk, and contingencies.
- Availability of technical resources.
- Well-defined communication plan between the project team and the client.
- Correct selection of ETL/EAI tool, if any.

Chapter 7
System Integration Testing

7.1 Objective

The purpose of the system integration test is to test the operation of the business system across and between application systems. Business system testing emphasizes on a common planning approach for all types of testing and advocates the reuse of deliverable components to test successively larger aspects of the application system.

The testing process begins with defining the testing requirements and strategy. After the testing requirements and strategy is prepared, testers begin developing unit, system, and system integration test scripts. These test scripts are used to guide testers through the various stages of the testing process. While test scripts are being prepared, one or more testing environments will be configured. The testers, to perform their unit, system, and system integration tests, use these testing environments. Once testing is completed in the test environment, users are supported in their acceptance testing of the new application system. For those environments where there are no custom extensions and interfaces to legacy or third-party systems, there is no need to perform unit and system integration testing.

The testing approach is integral to the *entire* implementation effort and is structured to build upon itself. The primary deliverables of the testing process are high-quality application systems that include both packaged applications and custom extensions (Fig. 7.1).

7.2 Business Flow Diagram

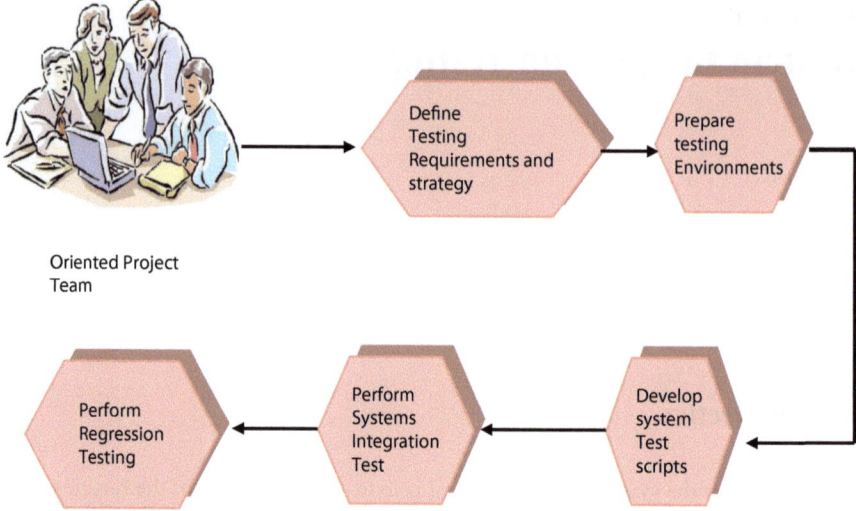

Fig. 7.1 Business flow diagram

Table 7.1 Activities for system and regression testing

Activities	Month 1				Month 2			
	1	2	3	4	5	6	7	8
Define Testing Requirement and Strategy	■							
Prepare Testing Environment			■					
Perform System Integration Testing					■	■		
Regression Testing							■	■

7.3 Implementation Schedule—System and Regression Testing (Table 7.1)

7.4 Prerequisite

- All application module setups are completed and setup design documents are updated
- Standard transactions are successfully performed in all application modules
- Coding for all custom components is completed and all custom components are unit tested
- Bugs identified during unit testing are fixed
- Design documents of custom components are updated

Table 7.2 Process, checkpoints, and weightage for testing requirement and strategy

Phase	System testing
Ref No	Task
1	Verify the testing description in the project-level scope, objectives, and approach contained within the project management plan
2	Identify the scope of the testing, types and iterations of tests, and relationships to other projects and systems
3	Identify the high-level approach for carrying out testing activities
4	Review relationships between process tasks/deliverables and other aspects of the overall project
5	Define the resource requirements for the testing activities of the project
6	Define integration test scripts and test cases

7.5 Detail Activities

7.5.1 Define Testing Requirement and Strategy

In this phase, identify the business system testing requirements and the strategy to be used for the testing of the system you are implementing. This also includes the recommended approach to testing, how to manage testing errors, and the criteria for accepting test results.

The testing requirements and strategy includes a listing of the types and the purpose of each testing task. In addition, it covers the tools used to perform testing. In the testing requirements and strategy, you establish or provide the following:

- A list of testing requirements
- An understanding of the type and the purpose of each testing task
- An understanding of the deliverables for each testing task
- An overview of the testing tools
- An overview of how problems will be managed
- Detailed acceptance criteria for testing

The objective of this task is to prepare a working strategy document that team members can refer to during the tasks in business system testing. New testing team members can quickly become familiar with the process by reviewing this document (Table 7.2).

7.5.2 Prepare Testing Environment

In this phase, install and configure one or more testing environments to support all testing activities. Configure the test environments based on the latest application setups from the application setup documents. Before beginning the testing in each environment, you should (Table 7.3):

- Set up the applications
- Decide on the real business data to be used

Table 7.3 System testing

Phase	System testing
Ref No	Task
1	Review system architecture strategy to understand the strategy for deployment of project environments and the testing environments in particular
2	List any other software applications needed to support business system testing
3	Update the testing tools component to reflect the configuration for each testing tool
4	Set up the testing environment
5	Install the application extension programs and testing tools

- Plan the volume of data (for example, converted data)
- Verify that all application parameters have been set up to enable transactions
- Verify that sufficient business data have been entered to demonstrate application features
- Provide access to definition and set up screens to appropriate users
- Enable links to non-enterprise resource planning (ERP) or custom systems (if applicable)
- Make reporting, data query, and testing tools available in the testing environments to verify correct results against the expected results of the test scripts

7.5.3 Perform System Integration Testing

At this point, test the integration of all business system flows within the target application system, including all standard and custom processes and reports. This task is equivalent to a full conference room pilot (CRP), where the environment simulates the future production environment. The system test is performed in a test environment.

It is important to test operating routines and procedures as well as programs, thereby simulating business operations, and not just systems. The repetitive nature of this task allows the project team to test each business flow and incrementally improve the system until they reach the desired level of success. Test each process flow until you achieve success. Then integrate the script with each other across process boundaries.

The supporting deliverable for this task is the test report for system test. This report includes a summary of the goals and objectives of the test, the business scope, a description of the test scenarios and processes, the final results, status, and recommendations. System test results are useful for implementing revisions to the business solutions. System testing process will also include

7.5 Detail Activities

Table 7.4 System integration testing

Phase	System testing
Ref No	Task
1	Review the responsibilities for the system test with the team
2	Validate the system test environment
3	Execute the integration system test script
4	Document the detailed test results

Table 7.5 System regression testing

Phase	System testing
Ref No	Task
1	Prepare regression testing scripts
2	Execute the regression test script
3	Document the detailed test results

Changes in acceptance criteria based on the new factors or changes in the business system (Table 7.4)

- Impact assessment of changing procedures and organization policy
- Actions and assignments relating to changes in procedures and policies
- Summary of retraining required supporting policy and procedure changes
- Highlights of critical failure in code
- Completed test scripts as an appendix
- A list of technical issues communicated and logged with ERP Support

7.5.4 Regression Testing

The regression test retest changes were made to a modified application extension because of an enhancement or a correction to a coding error. The regression test gets its name from the principle that when the development of a system progresses, validation of that progression requires proof that all prior validations have not regressed. In regression testing, retest application extensions are performed in order to validate that prior application extension defects have been corrected.

Regression testing needs to select test scripts that retest all of the application extensions that invoke the common module. Some projects use the regression test to stabilize application defects for the next iteration of the system test.

Regression test automation is now a very realistic option. Many tools feature the ability to record user actions and document graphical user interface (GUI) objects, thereby capturing this information into test scripts. You can then generalize the scripts to use variable data, enabling the script execution to be data driven. If your project is considering automating its regression testing, clearly state your automation goals and be sure to plan extra time for script development and generalization (Table 7.5).

Table 7.6 Deliverables and responsibilities

Sl No.	Name of deliverable	Template ID	Responsibility	
			Major	Assisted
1	Integration testing strategy document		Consulting organization	Client
2	Integration testing test cases/scripts		Consulting organization	Client
3	Regression testing test cases/scripts		Consulting organization	Client
4	Integration testing test results		Consulting organization	Client
5	Regression testing test results		Consulting organization	Client

Table 7.7 Major activities, checkpoints, and weightage

Major activities	Checkpoints	Weightage
Define testing requirement and strategy	Have all business scenarios been captured for all functional modules and custom extensions? Have all business test data been collected? Has testing been reviewed?	
Prepare testing environment	Are all recent patch sets applied in the Test Environment? Are all custom extensions migrated and registered in test environment? Are custom schema and database objects created in test environment?	
Perform system integration testing	Whether end-to-end business scenarios/cases have been prepared for all functional modules? Whether business data are being captured for integration testing?	
Regression testing	Whether the test scripts are prepared for the regression testing?	

7.6 Deliverables (Table 7.6)

7.7 Decision Matrix/Checklist (Table 7.7)

7.8 Critical Success Factors

The critical success factors of business system testing are as follows:

- A project plan that endorses an integrated testing approach; adequate resources and time for test script development and execution and support of testing environments
- Adequate tools, including properly configured environments and well-trained users, to support test script development and execution

7.8 Critical Success Factors

- Reliable, timely (converted) test data for each testing environment
- A thorough execution of all business system and system integration tests
- Independent QA testing and quality sign-off on all testing activities

The managers should be notified early that their staff (key users) will be involved in testing.

Chapter 8
Training

8.1 Objective

The skill transfer is an essential step in an enterprise resource planning (ERP) implementation project. The training is conducted at two levels for the client members in the project. In the first level, the key users/process owners are required to be trained on the functionality and major features of the software package. This training, if given before the business process analysis, helps to improve the communication between the consultants and the key users/process owners during the analysis phase. The next level of training is required for the end users where more stress is given on the transactions, reports/interface/other programs, user security, and system administration activities (applicable for the system administrator only). This training is required before the commencement of acceptance testing. Any other training may be arranged also, if required.

Successful projects require that end users be properly trained on the completed system. This step provides the requisite user procedures and training program to support end-user transition to the new system.

For each phase of training, the objective and the audience are to be known first for effective design of the training course.

8.2 Workflow Diagram of Training Process

The workflow of the training phase is shown in Fig. 8.1.

Fig. 8.1 Workflow diagram of training program

8.3 Implementation Schedule: Training (Table 8.1)

8.4 Prerequisite

- System integration process has been completed
- Organizational expectation from the training program is defined clearly
- Provisions made for the workload while project team members participate in the training program
- Early establishment of ongoing communication and feedback/evaluation mechanisms that fit the organizational environment
- Clear description of new job evaluation and grades

8.5 Detail Activity and Task List for Training

8.5.1 Assess User Requirements

Identify target audiences, define their documentation needs, define training requirements, and prioritize requirements.

8.5 Detail Activity and Task List for Training

Table 8.1 Implementation schedule: training

Activities	Month 1				Month 2				Month 3				
	1	2	3	4	5	6	7	8	9	10	11	12	13
Assess User Requirements	■	■											
Develop Training Strategy			■	■									
Develop Training Documentation					■	■							
Deliver Training and Assess progress							■	■					

Identify the target audience for the user documentation and determine their characteristics. This should also include management's needs. Answer questions such as:

- Who will be using the system?
- Where are the users located?
- What will be the frequency of updates both for the system and for the user documentation?
- Will there be different levels of users?
- What sorts of skills will each user level have?
- Is resistance anticipated from any user level?
- What are the organization's standards or preferences for user documentation delivery?
- Does the organization have development experience with online document systems?

Once the target audiences have been identified, define their documentation needs; for example, a user with little or no computer background would need documentation with a very different focus from a user with sophisticated systems skills. Each user procedure is associated with specific starting activities, implicit processes, and ending activities. Identify the boundaries of a user procedure and include the starting and ending points. These include:

- Manual processes
- Data stores
- Automated interfaces
- Report production
- Data entry

To develop appropriate training programs, information regarding the needs of each group of potential training students must be assessed:

- Level within the organization (e.g., management, clerical, operational)
- System responsibilities (e.g., users, system maintenance, operations)
- Experience level and skill base of identified personnel

Table 8.2 Assess user requirements

Process	Training
Ref no.	Task
1	Identify training objectives and contents
2	Identify target audiences for training
3	Estimate infrastructure and support requirements for training
4	Identify the boundaries of a user procedure and include the starting and ending points

Determine how each of the personnel within a group will interact with the new system. Each group will interact with the system differently (e.g., management may be enquiry-only users, while data entry staff may be online maintenance users only). Each of these different groups may require separate training courses that should be tailored to that group's specific needs. Gather information about target audiences including:

- Number and location of audience members
- Groupings and subgroupings
- General description of work responsibilities
- Position within the organization
- Work conditions
- Knowledge and skill levels
- Limitations
- Areas of exceptional performance
- General attitudes or culture

Compile the information gathered through interviews and observation and from existing documentation. Tabulate questionnaire and/or test results. Consolidate similar data and list target training audiences on the training delivery approach form. The numbers and locations of personnel to be trained in each group may affect the instructional method chosen, training resources required, number of training sessions scheduled, and/or required training time needed. Present the potential trainee list to the management to verify the assessment of those who must be trained and their relationships with the system. This information must be as accurate as possible as it will have an effect on the number, size, and type of training courses to be developed.

Note: In some countries, there may be legal restrictions on gathering personnel information, especially regarding attitudes and cultural background. Before this step is commenced, any law and statute compliance requirements should be assessed.

Assign a priority level to each training requirement (e.g., 1 = high, 2 = medium, 3 = low, 4 = N/A) to represent its contribution to achieve the organization's performance goals. Reorder the list from highest to lowest priority. Document each training requirement and its priority level on a training requirement form (Table 8.2).

8.5.2 Develop Training/User Procedures Strategy

Once end-user requirements are defined, create a strategy to develop end-user procedures documentation and create end-user training courses.

Once the user levels and document needs have been identified for the target audience, define the types of documents to be developed that are appropriate for the target audience. Paper-based document types may include:

- System overview/summary
- User manuals or guides
- User procedures
- Systems procedures
- Supplementary documents such as help desk procedures

Automated document types may include:

- Online help: electronic files that form an integral part of the application
- Information: electronic versions of user manuals that can be referenced with differing degrees of sophistication, depending upon the functionality of the access software used

Define or confirm documentation standards. These should address:

- Numbering and naming conventions
- Style (language and text format)
- Organization issues (e.g., chapters or modules)
- Format, stationery sizes and types, and binding preferences
- Word processing/graphics/desktop publishing software to be used for initial preparation and ongoing maintenance
- Production policies

Where the organization has existing user documentation standards in use, they should be adhered to or altered accordingly.

Determine how the documents are to be distributed, such as in three-ring notebooks, bound volumes, on diskettes, or on compact disc. Establish and agree upon the responsibilities for:

- Initial writing and creation
- Initial publishing
- Ongoing maintenance/updates
- Storage/distribution
- Training linkages

Prepare a narrative that describes the format of the documents. This should describe the organization and medium of the documents and include sample forms, a catalogue and description of icons (symbols) used, and a list of any reserved words and definitions. This task should also indicate how the materials are to be used, e.g., should the user read it, use it in concert with the application, or use it as a training reference.

Table 8.3 End-user training course

Process Ref no.	Training Task
1	Determine strategy to develop end-user procedures documentation
2	Determine type of document to be produced, online or paper based
3	Determine documentation standard
4	Develop a work plan to create the end-user procedural documentation
5	Confirm knowledge transfer plan/schedule

The proposed user documentation strategy is discussed with management and formal written approval is obtained. Resolve any questions or issues that arise.

Develop a work plan to create the end-user procedural documentation. Determine how training will be delivered:

- Third party (ERP education)
- Train the trainer
- Consulting organization
- Project team

The curricula design is driven by the training requirements derived in earlier steps of the method. Leverage the end-user procedures developed to determine what the training curricula should be. Testing and evaluation strategies are drafted and combined with the course outline from the curricula plan which delineates the specifications for effective training design and development.

When developing the training scope and strategy, be sure to consider all methods of training which can be provided (for example, consulting organization taught, train the trainer approach, third parties like ERP education, etc.). Document training scope and strategy and obtain client management approval of it (Table 8.3).

8.5.3 Develop Training Curriculum and End User Documentation

Develop the training course to teach end users their related procedures in the context of the new ERP system. Identify and select instructors for each course in the curricula plan. Instructors may be internal (train the trainer strategy), external, or a combination of both. Whichever strategy is employed, the identification of instructors is critical to assuring instructor capability and to define the overall training budget.

- Determine the instruction strategy for each course
- Develop standards for each type of training

8.5 Detail Activity and Task List for Training

Where an organization has existing standards for course materials and presentations, these should be adopted. If standards do not exist, they should be created for such items as:

- Numbering and naming conventions
- Style (language and text format)
- Organization issues
- Format, stationery, and binding
- Paper size
- Testing
- Responsibility for ongoing storage, maintenance, publishing, and distribution
- Production responsibilities

Standards are required for:

- Presentation and preparation software (e.g., word processing, graphics, desktop publishing)
- Student workbooks—style and content
- Overheads and 35 mm slides
- Instructor guides—style and content (optional)
- Computer-based training authoring software

Complete a structured walk-through of the documented standards with management for review. The development process should largely follow the steps established for the development of the help system. Some additional/different steps may be required where alternative software such as hypertext or computer-based training software is used as the development/presentation tool. For example, storyboarding techniques for computer-based training may be required for text development and presentation.

Identify end-user documentation to be produced based upon end-user documentation requirements that were identified earlier in this phase. User documentation should include:

1. Document introduction
2. Table of contents: Develop a structure to incorporate all user procedures identified into groups of logically related processes. Each logical grouping of user procedures is defined in the table of contents for each of the documents.
3. Documentation overview: Prepare the user documentation overview. The user documentation overview for each user procedure provides a visual representation of the system and the procedures needed to support the system. It shows the user's interactions with the system and defines the scope of the system to the user. Place all identified user procedures into groups of logically related processes. Define a numbering scheme for the user procedures and assign a number to each user procedure.
4. Management summary: Prepare the management summary, which is a general description of the system. It should incorporate an updated management overview diagram and support the user documentation overview by explaining the system in terms of:

a. Business needs and objectives that the system is designed to meet
b. Scope of the system in terms of the processes incorporated to meet the objectives
c. What the system does
d. Primary users of the system
e. Key features of the system
f. Major inputs and outputs of the system

Develop end-user procedures, which will be leveraged in end-user training.

Prepare the user procedures that form the main body of the document. They describe the manual activities to be executed in sequence. Their main purpose is to inform users on how to complete the different tasks that compose the various processes in the system and identify:

- The starting and ending points of the procedure
- The steps to be completed
- The sequence of the steps
- The person(s) responsible for completing the steps

The main components of each user procedure are:

- Responsibility: who completes the action
- Procedure steps: what action to take presented in a logical time sequence and any dependencies on other user procedures
- Supporting documents: which things (e.g., screens/windows, reports, input forms, or control logs) are related to and needed to complete the procedure step

Begin each procedure step with a strong action verb telling the person responsible what to do. Describe the procedure steps in the order of occurrence and ensure that they are sequentially numbered.

Prepare the detail task descriptions. Where the responsibility for a set of procedure steps is long, complex, or detailed, it may be easier to document these steps as a detail task description. These descriptions, which are written at a more detailed level than the user procedure, outline work steps for a single individual. They describe the sequence of work steps in a similar manner to the user procedure with action verbs and work steps in a logical time sequence, sequentially numbered.

Assemble and cross-reference all reports. Samples of each report are gathered together for inclusion with the user procedures to provide a visual representation of the information produced by the system. Cross-reference the reports to the procedures to which they relate for ease of use. The report layouts can be either a mock-up of the report or a copy of an actual report from the system. The means used to create or initiate the reports is also included.

Samples of each screen/window are gathered together for inclusion with the user procedures to provide a visual representation of the information produced by the system. The screen/window layouts are cross-referenced to the user procedures they relate to for ease of use. The screen/window layouts can be either a mock-up of the screen/window or a copy of an actual screen/window for the system.

Samples of each input form and input screen/window layout are gathered together for inclusion with the user procedures. These provide a visual representation

8.5 Detail Activity and Task List for Training

Table 8.4 Development of training course to teach end users

Process	Training
Ref no.	Task
1	Develop training course to teach end users
2	Develop training course to teach end users

of the information sent to the system. The input forms and input screen/window layouts are cross-referenced to the user procedures they relate to for each use. The Data Preparation Instructions document the name of each data element and provide a description of each data element.

Establish for each report generated by the system details of the numbers of copies to be produced and the distribution locations. Also establish details of retention and archiving requirements for each report. Ensure that the same approach is used to maintain transaction history or other data to be stored on tape or other storage media. Using the requirements of the types of documentation to be developed, which were produced earlier in the phase, prepare an introduction for each document. This should describe:

- What aspect of the system the document addresses (e.g., user guides, system operations manuals)
- Which users the particular document addresses (e.g., casual users, full-time users)
- The intended use of the document (Table 8.4)

8.5.4 Deliver Training

Based upon the courses and the user audience identified, develop a training schedule and assign end users to courses. Consider critical times in the client's calendar such as month-end close. Work with client management to develop the schedule to ensure that the training does not place undue stress on the client.

Obtain approval of the training schedule from client management.

Conduct training courses.

Prepare the training environment. The location of the ERP training environment can vary greatly depending on the approach and strategy. For instance, the training environment could exist in the ERP integration environment as a separate ERP client, or it could exist directly on its own independent ERP training system.

Confirm the ERP software installation strategy and requirements with project management, the appropriate business team leaders, and appropriate third-party support representatives.

Review the ERP-provided pre-installation system checklist. Work with ERP to ensure that the planned software delivery schedules are completed as scheduled and that ERP provides adequate information on the ERP installation. Ensure that all hardware, software, and networking components are in place and have been thoroughly tested.

Complete the ERP software installation. Work with certified ERP software installers to install the ERP software. Identify appropriate operations resources to participate in the installation process. Prior to installing the ERP software, the hardware supplier's operating system should already be installed and configured. Load the ERP software as follows (major steps only):

- Generate the UNIX or Windows NT file systems and operating system kernel
- Import the ERP software from tape/CD
- Install the database system (e.g., ERP, Informix)
- Install any third-party software
- Configure the application servers (if appropriate)
- Start the ERP system and ensure a successful log-on to ERP

Once the ERP software has been installed, complete a series of immediate ERP basic steps using the technical infrastructure team to access the system. These steps include:

- Create and configure the initial ERP clients
- Configure the initial ERP security access requirements
- Configure the ERP correction and transport system
- Configure the ERP Computer Center Management System (CCMS)
- Configure initial printer access capability
- Configure any optional high-availability requirements

Upon completion, obtain formal written approval from the technical infrastructure group indicating that the ERP system is ready for post-installation steps or project team access. Once the initial basis requirements have been completed, install the ERP software and complete any post-installation steps for the final preparation of the ERP system. This includes:

- Install and configure the ERP online CD-ROM. This is a Windows-based PC documentation product that can run locally on a PC or via a LAN server connected to PC workstations. This documentation is segmented into areas of accounting, logistics, human resources, and business process technology and basis.
- Complete a high-level cycle test using the business teams to verify that the application standard functionality is operating correctly. This should further ensure that the installation process has been executed correctly and is complete. This test should include a walk-through of specific major applications processes.
- Configure the Online Support Services (OSS) facility. OSS is ERP's service facility for obtaining current information concerning problems related to ERP standard software. Newer release versions of ERP now contain OSS as a process within the application rather than as a separate product.

Install any third-party software that could not be installed during the ERP software installation process.

Create the specific directories needed at the operating system level, e.g., UNIX directories used to store legacy data, which will later be read by an ERP program.

8.7 Decision Matrix/Checklist   119

Table 8.5 Delivery of training

Process Ref no.	Training Task
1	Prepare training environment
2	Train system administration
3	Conduct module-wise training session
4	Conduct module-wise hands-on sessions
5	Conduct case study session
6	Conduct exit test
7	Collect feedback of the training session
8	Determine readiness of business user group for UAT

Table 8.6 Deliverable, template ID, and responsibility

Sl no.	Name of the deliverable	Template ID	Responsibility	
			Major	Assisted
1	Training strategy		Consulting organization	Client
2	Training plan		Consulting organization	Client
3	Training handout		Consulting organization	Client
4	Conduct training		Consulting organization	Client
5	Training feedback		Consulting organization	Client

Once the training environment has been finalized, complete a structured walkthrough to ensure that the environment is ready for use. Make any corrections required.

Complete the training sessions according to the training program. Observe participant reaction to the course and request participant and instructor evaluation(s). If necessary, modify the training course based upon the evaluations. Ensure that responsibilities have been defined for ongoing presentation and maintenance of training courses and materials. As changes are made to the system, the training courses and materials need to be maintained to keep them in step with the system. In addition, ongoing training courses may be required during the life of the system to cater to new staff members or staff transferred from other areas. Define responsibilities for the ongoing maintenance, preparation, and delivery of the training materials and courses. Discuss and obtain approval. A balance must be drawn regarding timing of classes and the desire to conduct training close to point of start-up to assure "freshness" and the need for multiple iterations regarding complex subjects.

Evaluate and modify training approach and or user documentation as required (Table 8.5).

8.6 Deliverables (Table 8.6)

8.7 Decision Matrix/Checklist (Table 8.7)

Table 8.7 Activities, checkpoints, and weightage

Activities	Checkpoints	Weightage
Assess user requirements	Has list of system users been identified? Are all users located at the same place? Is the system and user documentation updated frequently? Will there be different levels of users? What sorts of skills will each user level have? Is resistance anticipated from any user level? Does the organization have standards or preferences for user documentation delivery? Does the organization have development experience with online document systems?	
Develop training strategy	Has the type of document appropriate for the target audience been decided (i.e., paper-based or automated training material)? Does paper-based document include: System overview/summary User manuals or guides User procedures Systems procedures Supplementary documents such as Help Desk procedures? Does automated document types include: On-line help: electronic files that form an integral part of the application Information: electronic versions of user manuals that can be referenced with differing degrees of sophistication, depending upon the functionality of the access software used? Has the documentation standard been decided? (These should address: Numbering and naming conventions Style (language and text format) Organization issues (e.g., chapters or modules) Format, stationery sizes and types and binding preferences Word processing/graphics/desktop publishing software to be used for initial preparation and ongoing maintenance Production policies) Has the client accepted training strategy?	
Develop training documentation	Has the documentation standard being followed in all handouts? Are all contents in end-user documentation in place? These should include document introduction, table of contents, documentation overview, management summary	
Deliver training and assess progress	Is the training delivery infrastructure in place? Is the O/S and ERP software installation complete? Is the timing and other syllabus communicated? Is the evaluation of training done? Any changes to training methodology/contents to be made based on the feedback?	

8.8 Critical Success Factors

- Availability of the organization's stakeholders to participate fully in readiness tasks
- Inclusion of key stakeholder constituencies in the change process and clear description of project impact
- Mechanism to listen and respond to top concerns about the new systems
- Management activities that help employees understand their new performance objectives and expectations, as well as the importance of their job to the change effort
- Commitment of management to post-performance support deployment measurement

Chapter 9
User Acceptance Test

9.1 Objective

This is the phase where the client validates the software to find out whether the solution provided is fit for their use and satisfies their business needs. After successful testing of each software component by the consultants themselves, it is handed over to the client for acceptance testing. Bugs/defects detected during acceptance testing are to be rectified accordingly before obtaining the certificate of acceptance from the client.

The client provides the acceptance test plan and test cases for testing.

The purpose of this phase is to support client business users in performing their acceptance test of the new enterprise resource planning (ERP) application system. The acceptance test is performed in the user acceptance testing (UAT) instance. This task also involves scheduling the acceptance test team, support staff, and user facilities.

The amount of acceptance testing is dependent on the specific acceptance criteria defined in the Unified Project Plan. The scope of the acceptance testing may change if client business users are participating substantially in the system and integration testing activities.

The tests are successful if they meet the acceptance criteria. Ideally, the acceptance criteria should match what the users think the system should do; however, the acceptance test should only be validated against predefined acceptance criteria, and not against what the users wish the new system would do.

The results of the acceptance test in the test report for the acceptance test should be recorded and communicated. This report should ideally contain:

1. Summary (including statistics) of successes of tests organized by the process.
2. Actions and assignments relating to changes to procedures and policies.
3. Highlights, action plan, assignment, and schedule for completion of critical failure in code.
4. Summary of repeating select business systems tests based on these changes.
5. Completed test scripts as an appendix.
6. A list of open technical issues that have been communicated to ERP support through a third assessment report (TAR).

Table 9.1 Activities and timeline

Activities	TimeLine												
	Month1				Month2					Month3			
	1	2	3	4	5	6	7	8	9	10	11	12	13
Prepare UAT Strategy		■											
Prepare UAT Environment			■										
Conduct UAT						■	■						
Signoff User Acceptance Certificate								■					

The acceptance test verifies that the new application system includes all of the required functionality.

It is to be noted that the consultants should not perform the acceptance test as it is ultimately the organization that must verify, through their testing efforts, that the new system meets the predefined acceptance criteria.

A successful acceptance test requires dedicated resources for conducting the acceptance test. Schedule the test far in advance so that managers will allow users the time off to perform their acceptance test (Table 9.1).

The consultants must also perform one or more dry runs with the UAT scripts before the business users start performing the tests for a better system performance and reduce issue resolution time (Fig. 9.1).

9.2 Implementation Schedule—UAT

9.3 Prerequisite

- **Project management plan**
  The Project Management Plan provides an overall strategy of testing for the project and an approach for acceptance of the test results.
- **Integration-tested system**
  Once the system's integration test is complete, the acceptance test begins.
- **Skilled users**
  Users receive training on the new system before they perform the acceptance test. The system and database administrators who know how to support the system, as well as designers who can provide functional and technical support for the tests, should also receive training.
- **UAT environment**
  The UAT environment needs to be fully configured and ready for the testers to begin their acceptance test. This includes platforms, software, test data, and documentation.

9.4 Detail Activities

Fig. 9.1 Business flow diagram

Table 9.2 Processes for user acceptance testing

Process	User acceptance testing
Ref no	Task
1	Prepare user acceptance testing plan (client activity)
2	Approve UAT strategy document

9.4 Detail Activities

9.4.1 UAT Strategy

The UAT strategy will consist of a detailed UAT execution plan. In this plan, determine the module-wise timeline when the UAT would be conducted and identify the business users who actually perform the UAT. Identified business users will prepare the test script and test cases for UAT. It is advisable to conduct UAT after month-end or quarter-end closing, as business users are normally tied up with regular month-end and quarter-end close activities during this period. Once the strategy is formulated, the same should be discussed in the steering committee meeting to get approval from the client side. The approved UAT strategy document will be the baseline for the execution of UAT (Table 9.2).

9.4.2 Prepare UAT Environment

A separate instance should be created for UAT and this instance should have all the software required for conducting UAT. All the business users, who will perform UAT, should have the required client machine software (e.g., Application Development Interface (ADI) Initiator, etc.) in their respective desktops. Business users will create the business scenario and test cases for UAT. Test cases and test plan should be exhaustive enough so that all the functionalities of the system can be tested during UAT (Table 9.3).

Table 9.3 Processes for the preparation of UAT environment

Process Ref no.	User acceptance testing Task
1	Create business scenario (client activity)
2	Create UAT data for testing transaction
3	Prepare UAT environment/instance
4	Prepare UAT test cases and test plan

Table 9.4 Processes for conducting UAT

Process Ref no.	User acceptance testing Task
1	Perform test against test cases
2	Update/modify ERP application setup and custom programs
3	Update/modify set-up document and custom component design document
4	User acceptance sign off

9.4.3 Conduct UAT

Business users will perform UAT and the consulting organization will provide all the necessary support in conducting the UAT. If any bug is identified, it will be fixed and simultaneously the setup document/technical design document and program will be updated accordingly. Configuration management procedure should be strictly adhered to for any changes in the setup document/technical design document. An UAT summary report is to be prepared to consolidate the results of UAT. Finally, when all the components are passed through UAT, the formal UAT sign-off certificate should be collected from the client (Table 9.4–9.6).

9.5 Deliverables

9.6 Decision Matrix/Checklist

9.7 Critical Success Factors

- A separate UAT instance is created and the entire required patch set is applied.
- Business users create UAT script before start of integration testing so that all the test scripts get tested at the time of integration testing.
- The same business users, who will conduct UAT, should create the UAT test script/test cases.

9.7 Critical Success Factors

Table 9.5 Deliverable, template ID, and responsibility

Sl no.	Name of the deliverable	Template ID	Responsibility	
			Major	Assisted
1	UAT environment		Consulting organization	Client
2	UAT test plan and test cases		Consulting organization	Client
3	UAT summary report		Consulting organization	Client
4	Acceptance certificate		Consulting organization	Client

Table 9.6 Major activities, checkpoint, and weightage

Major activities	Checkpoint	Weightage
Conduct UAT	Is the UAT execution complete?	
	Has the UAT summary report been prepared and accepted by the client?	
	Has open technical issues/bugs referred to ERP support and entered in issue/TAR log?	
	Has the acceptance certificate been signed off by client?	

- All the business users should be fully available in the whole period of UAT so that UAT can be completed as per schedule.
- Business users training should be conducted before UAT and needs to ensure that business users are quite confident in using the ERP application.

Chapter 10
Production Go Live

10.1 Objective

The objective of production migration is to migrate the organization, systems, and people to the new enterprise system. The production environment is prepared in this phase. The database instances are created, applications are set up, and initial data (master data, open transaction, and required historical transactions) are migrated from the legacy systems to the new enterprise resource planning (ERP) applications. The system is audited and performance check is carried out. The system is assessed again after the Go Live and necessary measures are taken.

10.2 Workflow of the Production Go Live phase (Fig. 10.1; Table 10.1)

10.3 Prerequisite

- System/integration testing process has been completed.
- Training has been conducted for business users.
- UAT has been completed for standard ERP application processes and all custom components.
- Organizational and process changes are accepted and supported by all levels of the organization and senior-level managers.
- Demonstration is carried out by users that they can perform their jobs using the new documentation, tools, and systems.
- All application modules function are as expected in support of business requirements.
- Effective structure is to be in place to provide timely response to user queries and requests.

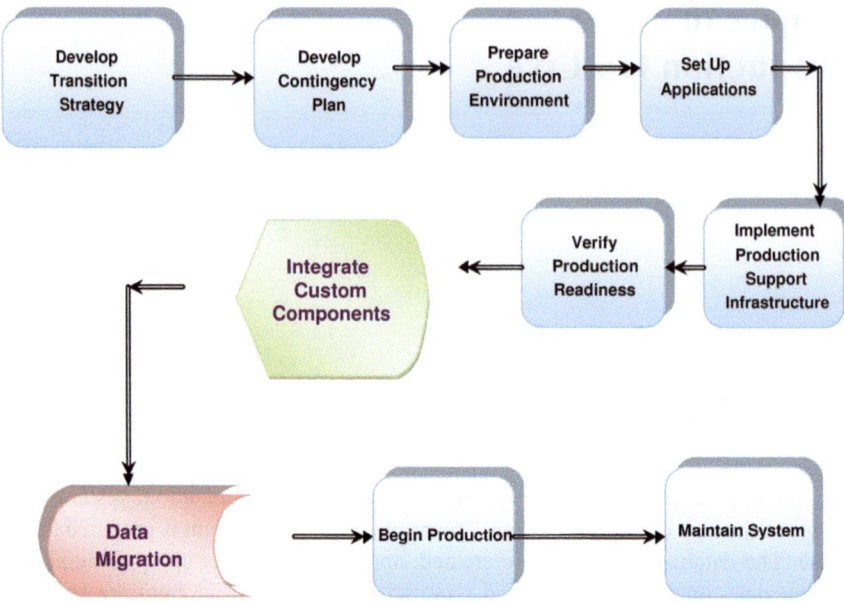

Fig. 10.1 Work flow of the production Go Live phase. Implementation schedule: production Go Live phase

Table 10.1 Activities for the production Go Live phase

	Month 1				Month 2				Month 3					Month 4				Month 5			
Activities	1	2	3	4	5	6	7	8	9	10	11	12	13	14	15	16	17	18	19	20	21
Define Transition Strategy																					
Develop Contingency Plan																					
Prepare Production Environment																					
Set up Applications																					
Implement Production Support Infrastructure																					
Verify Production Readiness																					
Integrate Custom Components																					
Data Migration																					
Begin Production																					
Maintain System																					

Table 10.2 Determining time window for cutover

Process	Production Go Live
Ref no	Task
1	Prepare detail production Go Live plan
2	Review the proposed system architecture, locations, and resources within locations
3	Assessing readiness for transition to production
4	Determine cutover date
5	Review the draft deliverable with senior management and seek approval

10.4 Detailed Activities

10.4.1 Define Transition Strategy

In this task, outline the business transition approach for migrating the system, organization, and people-to-production status. In order to complete the transition strategy, the following points are important:

1. **Determine Cutover Approach** The two basic choices here are to either perform a clean cutover from existing legacy systems to the new systems or to run the systems in parallel for a period of time, maintaining the existing system as a backup, if needed. This question is not always so clear cut, since there may be a desire for a clean switch to the new systems, but the existing system may need to be kept online to maintain existing active data that will not be converted into the new system due to cost or time constraints. This would require users to enter transactions in the new system and the old system. There may also be issues in reporting or consolidating data across the systems.

2. **Determine When Cutover Should Occur** The timing of the cutover to the new systems is usually determined early in a project as cutover is one of the key milestones in any high-level project plan. The precise timing is determined by the competing factors of the desire to cut over as soon as possible, thereby minimizing disruption of business operations and minimizing the need to perform cross-system transactions, reporting, and consolidation. The cutover milestone is usually at the end of the financial quarter or year.

3. **Determine Time Window for Cutover** The allowable time window for the production cutover is critical in determining what is or is not feasible and may influence the transition strategy profoundly. The time window usually needs to be kept to a minimum while considering the amount of risk that the business is willing to assume (Table 10.2).

10.4.2 Develop Contingency Plan

In this task, develop the detailed transition plan for moving onto the production system, as well as an implementation contingency plan. Planning for contingencies

Table 10.3 Developing contingency plan

Process	Production Go Live
Ref no	Task
1	Identification of the risks and failure points in the transition strategy
2	Outline a high-level course of action for handling each eventuality

in the migration process requires, as a first step, identification of the risks and failure points in the process. Having identified these areas, outline a high-level course of action for handling each eventuality. Because of the focus on risks and failure points, this task is tightly linked to verify production readiness. The detailed analysis that is used in verify production readiness can also be used here, at a higher level, for identifying general risk areas. The identification of contingency measures may be unpopular because it has connotations of project failure and mishandling, but a more thorough contingency plan can potentially reduce the amount and degree of crisis management needed, should the unthinkable happen. Do not confuse implementation contingency with system contingency failure (Table 10.3).

10.4.3 Prepare Production Environment

In this task, set up, configure, and install the database and application software for the production environment.

1. Technical Architecture In projects where there are complex hardware and software requirements, designing and implementing the architecture begins during application and technical architecture. The implementation of the production environment is based on the assumption that the system architecture has been created and all necessary preparation for the installation of the applications and custom modules has been performed.

2. Timing Considerations in Implementing the Production Environment Although prepare production environment is part of production migration, the installation of the production environment may have taken place earlier in the implementation when other working environments were installed. This is common in cases where management selected, purchased, or sized the hardware prior to or during the database and application selection process. It is also often true for smaller organizations that can only provide a single machine to support all project environments. Assuming that the hardware configuration is adequate to support production requirements, this task then serves as verification that the production environment is ready for production and that any custom modules have been installed. If the adequacy of the hardware configuration is in question, performance testing should be considered to identify performance risks and alternatives. If performance testing has already been performed, portions of the production environment may have been installed and configured to support this test. This task then serves as an opportunity to apply any configuration changes necessary to reflect recommendations

10.4 Detailed Activities   133

Table 10.4 Implementing the production environment

Process	Production Go Live
Ref no	Task
1	Review the system architecture solution design
2	Install the operating system and networking software on the servers
3	Install ERP application software on the server
4	Create custom database schema and custom objects
5	Create user accounts on the servers
6	Install printers and other peripheral devices
7	Verify the production system installation

made in the performance test report. This task represents the culmination of many tasks that contributed to the final application and technical architecture that must be implemented in the production system. The activities that comprise this process may span a wide time frame and be performed in parallel with tasks in other processes (Table 10.4).

10.4.4 Set Up Applications

In this task, implement the required setups in all of the application modules as part of configuration. Manually enter the application setups into the production applications database. The application setup documents produced during conference room pilot phase serve as a guide for entering setup data and define the correct setups for each application. Although you already have the application setup recorded, you should also capture the production setup after the data have been entered. This lets you compare the setup from expected to actual as a form of quality assurance. Use screenshots or standard reports as a means to make a record of production setup. If you discover that some data should be altered in production, then check with the key process or module owners to determine whether a slight change in setup would impact other operations. In this situation, the applications setup task belongs to the business analysts who have been most involved in the definition of applications setups throughout the life of the project (Table 10.5).

10.4.5 Implement Production Support Infrastructure

In this task, activate the support personnel and procedures for the new business system and review requirements for support-related services from software suppliers, contractors, help desks, and other support services. It is important to rehearse the support procedures. You can simulate various types of support calls and evaluate whether the expected response time, accuracy of the resolutions, and the general flow of the process are adequate to support the request volume. This is also a

Table 10.5 Setting up of applications

Process Ref no	Production Go Live Task
1	Create responsibility only for super user for respective modules
2	Configure major setup required before multi org conversion
3	Multi org conversion
4	Identify all the patch sets/patch to be applied on production instance
5	Review the patch list
6	Apply all the required patch set to production instance
7	Common application setup
8	Functional module setup
9	MRC conversion
10	MRC setup
11	Create all business users
12	Review production setup configuration

Table 10.6 Implementation of the production support infrastructure

Process Ref no	Production Go Live Task
1	Introduce the help desk to the organization
2	Set up and distribute reference materials
3	Implement the online help text
4	Implement the online issue tracking system
5	Establish a change request and bug-reporting procedure
6	Review procedures for getting help

good opportunity to test default support mechanisms, after-business-hours support, and supplier support hotlines. The establishment of a robust and capable internal and vendor external support infrastructure is critical to the success of your project (Table 10.6).

10.4.6 Verify Production Readiness

This is the final approval step before the cutover to the new production system.

1. User Readiness For all business processes, users should demonstrate that they are qualified to fulfill the requirements of the process as well as each downstream process step. Presentations by users, also known as *walkthroughs,* can be used to demonstrate competency levels. Finally, users should confirm that they feel prepared and confident to move to the new system. As part of the overall communication program, contact partners and suppliers have to let them know when production cutover will occur. Ask for additional support coverage for a set period of time. The organization management may also want to notify its key customers of the

10.4 Detailed Activities

Table 10.7 Processes for plan verification

Process	Production Go Live
Ref no	Task
1	Verify the completion of production environment
2	Confirm senior management commitment via the steering
3	Committee
4	Verify that users are trained
5	Confirm the production cutover

Table 10.8 Integration of the custom components

Process	Production Go Live
Ref no	Task
1	Migrate all custom forms
2	Migrate custom reports
3	Migrate custom workflow and alerts
4	Integrate custom interfaces
5	Migrate all custom objects (table view, packages, programs)

transition to a new system. In this way, they can be prepared to deal with any issues, potential delays in service, or changes to procedures or business documents.

2. **Production System Readiness** As another means to confirm that the system is stable, key users should go through the production environment setup and data to confirm that every form can be accessed and all reports are printed successfully. It is also a good opportunity to test user menus and verify that each user has the appropriate security access and responsibilities and that their user profiles are set up correctly. Verify that all business system test actions are closed.

3. **Contingency Plan Verification** The transition and contingency plan implementing the production system was created prior to beginning the transition to production. It is important to review the cutover contingency plan component with the information systems organization and, in addition, communicate the essence of the fallback plan to the users. In this case, if there is a production delay, an organization-wide announcement directing users to the cutover contingency plan can be made without having to provide detailed explanations (Table 10.7).

10.4.7 Integrate Custom Components

All custom components are integrated with the standard ERP application product suite. Custom components can be forms, reports, workflow, programs, and interfaces. Prior to this task, all custom components should pass through user acceptance testing and final version of all the components should be frozen in the test environment. These components then are to be integrated to base application in production

Table 10.9 Processes for data migration

Process Ref no	Production Go Live Task
1	Migrate legacy master data (e.g., vendor, customer, item, etc.)
2	Migrate legacy transaction data (open invoice, open receipts, open GL balances)
3	Migrate backlog transaction data

environment through a third-party tool (e.g., Kintana) or manually. Registration of custom components in ERP application is done in this task (Table 10.8).

10.4.8 Data Migration

Before going live in the production environment, the required master data, the open transaction data, and the historical transaction data need to be imported from the old legacy applications to ERP applications. Since data structure and data design in legacy systems are different from those of ERP applications, data need to be messaged/converted satisfying the business rules to the ERP requirement. Initial data can be migrated either by manual entry (where data volume is less) or from spreadsheet (where data volume is high) using some ERP/third-party tool (usage of tools are limited to certain ERP modules only) or by using some custom-built data migration program (for high and very high volume of data). ERP provides some standard open interfaces to receive data from the external systems. Data migration programs basically extract data from the legacy application and convert them to ERP supplied form. In case ERP standard open interfaces are found not suitable, new custom open interface needs to be developed. The activities for development of data migration programs are similar to those of interface development as described in the earlier phase (Table 10.9).

10.4.9 Begin Production

In this task, you confirm that organization-wide use of all aspects of the production system is in place. This task covers a period of time during which the following conditions apply:

- The environment is carefully controlled.
- Transactions are brought online in a priority sequence (most critical first) to keep support requirements low and focused.

The best way to begin using the new system is to control the initial transactions and entries. Bring up departments or sets of related transactions in sequence to permit focused attention on limited areas until they are stabilized. Use the production schedule. You must anticipate issues and quickly remedy them. Priorities may shift during the production state and therefore you must set expectations with users on what types of problems to resolve first (Table 10.10).

10.4 Detailed Activities 137

Table 10.10 Processes for initiation

Process Ref no	Production Go Live Task
1	Initiate using the production system
2	Initiate incident/technical issue management procedures
3	Initiate support
4	Confirm that all components of the production system are operational
5	Declare the new system live

Table 10.11 Processes for maintaining the system

Process Ref no	Production Go Live Task
1	Perform regular backups
2	Periodically defragment table spaces
3	Monitor and respond to database growth
4	Purge report and system log files
5	Apply software patches
6	Perform regular version upgrades
7	Implement user-driven changes
8	Update the system configuration document to reflect architectural changes

10.4.10 Maintain System

Both routine and on-demand activities are carried out at this point.
Routine maintenance includes:

- Running batch reports
- Running extracts
- Logging file purges
- Enlarging table space
- Monitoring table space
- Running hot backups
- Running cold backups
- Performing security audits
- Performance tuning
- Archiving and purging data

On-demand maintenance includes unplanned or requested activities such as the following (Table 10.11):

- Setting up new users
- Changing new user access
- Correcting interface data
- Correcting data due to user error

Table 10.12 Deliverables and responsibilities

Sl no	Name of deliverable	Template ID	Responsibility	
			Major	Assisted
1	Detailed transition and contingency plan		Consulting organization	Client
2	Helpdesk/support procedures		Consulting organization	Client
3	Detail data migration plan		Consulting organization	Client

10.5 Deliverables (Table 10.12)

10.6 Decision Matrix/Checklist (Table 10.13)

10.7 Critical Success Factors

The critical success factors of production migration are as follows:

- Organizational and process changes accepted and supported by all levels of the organization and senior-level managers.
- Demonstration by users that they can perform their jobs using the new documentation, tools, and systems.
- Data converted accurately and within the planned time frame.
- Installed and properly tested production hardware and software.
- All application modules function as expected in support of business requirements.
- Agreement by all involved parties on the transition schedule.
- Proven measurable performance objectives for achieving high levels of service.
- Effective structure in place to provide timely response to user queries and requests.

10.7 Critical Success Factors

Table 10.13 Major activities and checkpoints

Major activities	Checkpoints	Weightage
Define transition strategy	*Is the exact time of cutover defined?* The cutover milestone is usually at the end of financial quarter or year	
	Is the time window for cutover accepted by business? The time window usually needs to be kept to a minimum while considering the amount of risk that the business is willing to assume	
	Are all resources available for production Go Live? *Environment*: host system, network, communications, personal computers, and peripherals *Software*: version, compatibility, and integration of custom extensions *Users*: support, users, project team, management, and external staff	
Develop contingency plan	Will the current system still be available and fully operational?	
	Will data be lost or damaged if the organization is forced to implement the contingency plan?	
	Is the maximum amount of time for the contingency plan determined, which could be used, without jeopardizing the organization situation?	
	Is there a manual backup procedure in place?	
	Are users trained to recognize contingency situations and activate corrective actions?	
Prepare production environment	Does the business user's list receive from the business?	
	Does the detail list of printers receive from the business?	
	Is the number of nodes of production and test instances same? This is important for future cloning of test instance from production instance	
Setup applications	Has patch list been reviewed and correct patch set applied in the production environment?	
	Are all setup data available?	
	Are all setup documents updated?	

Table 10.13 (continued)

Major activities	Checkpoints	Weightage
Implement production support infrastructure	Is the communication tree in place?	
	Do different user mail groups exist for fast communication?	
	Is the issue escalation process defined?	
Verify production readiness	Is there approval from senior management for final cutover date?	
	Have all business users and concerned persons been intimated regarding the production Go Live downtime?	
Integrate custom components	Are you using any version control tool to maintain versions of custom components?	
	Have you moved the latest version of custom components to production environment?	
Data migration	Have you frozen all source files to be migrated?	
	Have you validated all sources by the clients?	
	Have you taken sign off from the client on detail data migration plan?	
	Have you prepared the sequence of data loading?	
	Have you ensured all backlog transaction data been loaded before the user starts using the system?	
Begin production	Do all the system users have proper training?	
	Do all the business users have correct responsibilities and menus attached to his/her user profile?	
Maintain system	Is disk utilization being monitored continuously?	
	Are different concurrent managers being created for all the programs, which take longer time to execute?	

Chapter 11
Rollout

11.1 Objective

While implementing enterprise resource planning (ERP) applications in multisite organization environment, it is implemented first at one pilot site. After stabilization of the system at the pilot site, the system is implemented at other sites using the procedures and configuration data, and customization of the pilot site is carried out as far as possible. The rollout strategy is formulated first. Then the difference in the requirement of the business processes, interfaces, data migration, and other technical aspects of the respective sites with respect to the pilot site are studied and defined. The next step is to update the corresponding documents based on the requirement definitions. Setups of the application package are changed accordingly. Customization, interface, and data migration programs are also modified or new programs are added and validated. Go-Live operations at rollout sites are similar to those of pilot sites.

Different project teams are required for each rollout site. The major stress of the rollout project team is to reuse the implementation process, set up data, technical and operational architecture, customized components, interface, and data migration programs as much as possible. Too much deviation from the pilot site implementation results in the delay in commissioning of the rollout sites and eventually increases the stress on the project budget.

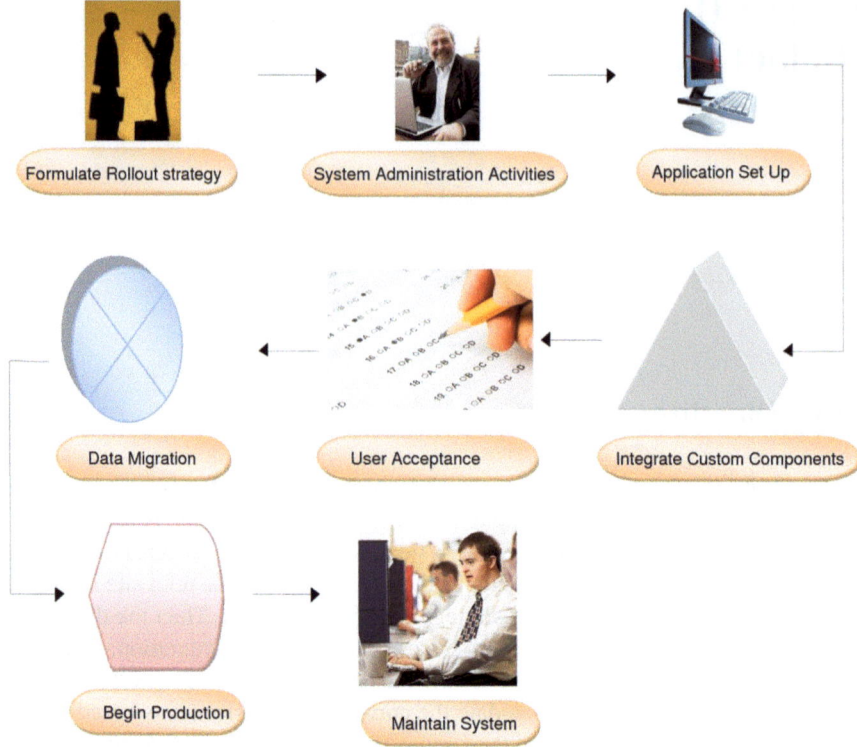

Fig. 11.1 Business flow diagram

11.2 Business Flow Diagram (Fig. 11.1)

11.3 Implementation Schedule: Rollout (Table 11.1)

11.4 Prerequisites

- Pilot site has undergone Go-Live successfully.
- All application modules function as expected in support of business requirements in the pilot site.
- All setup templates of pilot sites are updated.
- All functional and technical design specifications of custom components are updated.
- Clear scope definition has been given as a means to carry out rollout in multiple sites.
- Sufficient quality information about current and future business volumes to enable the definition of rollout plan is made available.

11.5 Detailed Activities

Table 11.1 Activities and duration in the implementation schedule

Activities	Month 1				Month 2				Month 3					Month 4			
	1	2	3	4	5	6	7	8	9	10	11	12	13	14	15	16	17
Roll Out strategy	■																
Analyze site specific requirement		■	■	■													
System administration activities					■												
Application set up						■	■	■	■								
Integrate Custom Components										■	■						
User acceptance Testing												■					
Data migration														■	■		
Begin Production																■	
Maintain system																	■

11.5 Detailed Activities

11.5.1 Rollout Strategy (Table 11.2)

11.5.2 Analyze Site-Specific Requirement

Global ERP application setup template would be created as part of the conference room pilot process. The same global template would be reused for multiple site rollouts. It is assumed that 70 % of global template will be further used in other rollout sites. Site-specific localization needs to be identified at this time. Customization is restricted to the preparation of maximum ten site-specific statutory reports. If other customization needs are identified during the analysis process, then timeline will be increased accordingly (Table 11.3).

Table 11.2 Rollout strategy

Process Ref no.	Rollout Task
1	Identify sites for rollout
2	Organize global rollout team
3	Organize resources for each rollout site
4	Prepare detailed rollout plan for each site

Table 11.3 Processes for analyzing site-specific requirements

Process Ref no.	Rollout Task
1	Review global setup template in the line of site-specific need
2	Determine localization need of each rollout sites
3	Determine localization patch set required for each rollout site
4	Identify at a high level reporting requirements of each rollout site
5	Prepare site-specific setup template
6	Prepare design document of site-specific custom components

Table 11.4 Processes for the preparation of production environment

Process Ref no.	Rollout Task
1	Review the system architecture solution design
2	Install the operating system and networking software on the servers (if multiple instances need to be created)
3	Install ERP application software on the server (if multiple instances need to be created)
4	Create custom database schema and custom objects
5	Create user accounts on the servers
6	Install printers and other peripheral devices
7	Verify the production system installation

11.5.3 *Prepare Production Environment (Table 11.4)*

11.5.4 *Application Setup (Table 11.5)*

11.5.5 *Integrate Custom Components (Table 11.6)*

11.7 Decision Matrix/Checklist

Table 11.5 Processes for application setup

Process Ref no.	Rollout Task
1	Identify all the patch sets/patch to be applied on production instance
2	Apply all the required patch set to production instance
3	Create responsibility for business user for respective modules
4	Common application setup
5	Functional module setup
6	Review production setup configuration
7	Build and unit testing of custom component

Table 11.6 Processes for integration of custom components

Process Ref no.	Rollout Task
1	Migrate all custom objects (table view, packages, programs, etc.)
2	Migrate all custom forms
3	Migrate custom reports
4	Migrate custom workflow and alerts
5	Integrate custom interfaces

11.5.6 User Acceptance Testing (Table 11.7)

11.5.7 Data Migration (Table 11.8)

11.5.8 Begin Production (Table 11.9)

11.5.9 Maintain System (Table 11.10)

11.6 Deliverables (Table 11.11)

11.7 Decision Matrix/Checklist (Table 11.12)

Table 11.7 Processes for UAT

Process Ref no.	Rollout Task
1	Define business/transaction scenario and test data (client activity)
2	Perform testing (client activity)
3	Assist in testing
4	Update setup document
5	Analyze issues and problems
6	Resolve issues and develop customization specifications

Table 11.8 Processes for data migration

Process Ref no.	Rollout Task
1	Migrate legacy master data (e.g., vendor, customer, item, etc.)
2	Migrate legacy transaction data (open invoice, open receipts, open GL balances)
3	Migrate backlog transaction data

Table 11.9 Processes for initiating production

Process Ref no.	Rollout Task
1	Initiate using the production system
2	Initiate incident/technical issue management procedures
3	Initiate support
4	Confirm that all components of the production system are operational
5	Declare the new system live

11.7 Decision Matrix/Checklist

Table 11.10 Processes for maintaining the system

Process Ref no.	Rollout Task
1	Perform regular backups
2	Periodically defragment table spaces
3	Monitor and respond to database growth
4	Purge report and system log files
5	Apply software patches
6	Perform regular version upgrades
7	Implement user-driven changes
8	Update the system configuration document to reflect architectural changes

Table 11.11 Deliverables and responsibilities

Sl no.	Name of deliverable	Template ID	Responsibility	
			Major	Assisted
1	Rollout strategy		Consulting organization	Client
2	Site-specific setup templates		Consulting organization	Client
3	Design specification for custom reports		Consulting organization	Client
4	Source code of custom component		Consulting organization	Client
5	Configured ERP application instance		Consulting organization	Client

Table 11.12 Major activities, checkpoints, and weightage

Major activities	Checkpoints	Weightage
Rollout strategy	Have the sites for rollout been identified?	
	Has priority been assigned to different sites for rollout?	
	Has the resource allocation been decided for the rollout sequence?	
	Are there enough resources for allocation for rollout?	
	Are the work plans, sequence of activities, dependencies, schedule, estimated time, and resource requirements for rollout ready?	
System administration activities	What will be the operating system for the rollout sites?	
	Has the database for each site been identified?	
	Are the configurations needed to support the solution decided?	
	Has the user's list from the business been received?	
	Have the number of users and their responsibility for the setup been checked?	
	Have the custom database schemas and custom objects been installed for using the system?	
Application setup	Are all required software and hardware available?	
	Is the setup environment validated for the configuration of the system?	
	Is the sequence of steps for setup been finalized?	
	Are all setup data available?	
	Whether the setup document is updated?	
	Are all patch sets and patches required for production instance been applied?	
Integrate custom components	Is there any custom component? If any, what are those components (list of custom components)?	
	Is the custom component migrated fully?	
User acceptance testing	Have the business scenarios to be tested been identified?	
	Are the transactions to be performed for each business scenario listed?	
	Are the test cases for each scenario and expected result from the test been found out?	
	Have the assisting testing plans, test case, and testing activity been properly planned out?	
	Are setup documents based on testing result updated?	
	Are the test problems and solutions to resolve the problem been identified?	
	Are there any requirements for customization to solve the above-mentioned problems?	

11.8 Critical Success Factors

Table 11.12 (continued)

Major activities	Checkpoints	Weightage
Data migration	Are the sources of files from legacy data required for migration ready?	
	Has the mapping of the external system data to migrate into interface table been done?	
	Does the client validate the source data?	
	Has the data migration strategy been planned with the user and sign-off taken?	
	Have the data loading sequences been finalized?	
	Are all the data loaded before the user starts using the system documented?	
	Is there any remaining data loading activity?	
Begin production	What is the level of confidence the user has in using the new system?	
	Have all the responsibilities been correctly assigned to individual users?	
	Has the functionality and performance measurement process been prepared?	
Maintain system	Is the database monitoring schedule being prepared?	
	Is the database performance being monitored on a regular interval?	
	Is the list of concurrent programs, which are taking longer time, been identified?	
	Do all business users have the correct responsibilities and menus attached to their user profiles?	
	Is the problem-tracking system in place?	
	Are all the patch sets applied in the operating system level been tracked?	
	Are all the patch sets applied in the ERP application system been tracked?	

11.8 Critical Success Factors

- Organizational and process changes accepted and supported by all levels of the organization and senior-level managers.
- Demonstration by users that they can perform their jobs using the new documentation, tools, and systems.
- Data converted accurately and within the planned time frame.
- Installed and properly tested production hardware and software.
- All application modules function as expected in support of business requirements.
- Agreement by all involved parties on the transition schedule.
- Proven measurable performance objectives for achieving high levels of service.
- Effective structure in place to provide timely response to user queries and requests.
- Availability of technical experts to tune the production environment.

Chapter 12
Project Management

12.1 Objective

The objective of project management is to provide a framework in which all types of enterprise resource planning (ERP) application projects can be planned, estimated, controlled, and completed in a consistent manner.

ERP application projects are characterized by a high degree of uncertainty.

The project management concept focuses on the additional discipline needed to ensure that client expectations are clearly defined at the outset of the project and remain visible throughout the project life cycle. Project management also formalizes control mechanisms to help the project team share critical project information and coordinate with external stakeholders.

The overall organization of project management is expressed as a process-based methodology, which can be tailored to a project's specific needs.

The five management processes are:

1. Control and reporting
2. Work management
3. Resource management
4. Quality management
5. Configuration management

The project management tasks can be divided under the following five categories which constitute the project management life cycle (Fig. 12.1):

1. Project planning
2. Phase planning
3. Phase control
4. Phase completion
5. Project completion

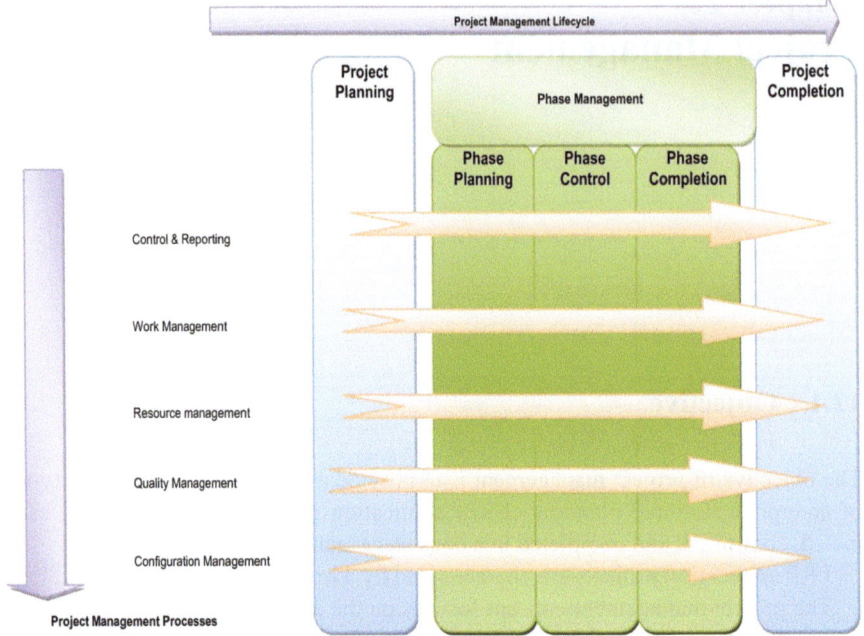

Fig. 12.1 Project management life cycle

12.2 Project Management Life Cycle

12.2.1 Prerequisites

Project Planning

- Client statement of work
- Consultant's proposal
- Contractual agreement
- Client policy document
- Consultant policy documents
- Resource cost estimates

Project Control

- Unified Project Plan: Refer section "Project Initiation"
- Project staffing and organization plan: Refer section "Project Initiation"
- Project orientation guide: Refer section "Project Initiation"

12.2 Project Management Life Cycle

Table 12.1 Processes in project planning

Process Ref No.	Project management Task
1	Establish scope, objectives, and approach
2	Define control and reporting strategies, standards and procedures
3	Establish schedule and work plan
4	Estimate effort and finance plan
5	Define risk management strategies and mitigation plans
6	Establish staffing and organization plan
7	Establish infrastructure plan
8	Define quality management strategies, standards, and procedures
9	Define configuration management strategies, standards, and procedures

Project Completion

- Unified project plan
- Progress reports
- Risk and issue log
- Change request log
- Audit action and reports
- Configuration management repository

12.2.2 Detailed Activities

Project Planning

Tasks in this category encompass the definition of the project with respect to scope, quality, time, and cost. Project planning tasks also determine the appropriate organization of resources and responsibilities to conduct the project (Table 12.1).

Please refer to the section on "Project Initiation" for a detailed description of the project planning activity and associated tasks and deliverables.

Project Control

The objectives of project control are to:

- Manage the scope, quality, cost, and schedule of phase tasks and deliverables to meet or exceed client expectations.
- Compare phase execution progresses to plans, identify variances, and adjust to correct significant variances.
- Anticipate possible risks to the project and take preventive measures to contain them.

Table 12.2 Processes in project control

Process Ref No.	Project management Task
1	Risk management
2	Issue/problem management
3	Scope/change control
4	Status monitoring and reporting
5	Work plan control
6	Effort/cost control
7	Physical resource control
8	Quality review
9	Quality audit and quality measurement
10	Support health check
11	Configuration control
12	Document control
13	Knowledge management
14	Release management

- Assure the integrity of phase deliverables and that changes to them are consistent with project objectives.
- Efficiently resolve issues and problems as they are identified, identify root causes, and take corrective actions.
- Develop and direct project staff to achieve a rewarding work environment and an efficient motivated project organization (Table 12.2).

12.2.3 Risk Management: Broad Approach

12.2.4 Issue/Problem Management: Broad Approach

Problem solving is the result of problem definition and decision-making. Problem definition requires distinguishing between causes and symptoms. Decision-making includes analyzing the problem to identify viable solutions, making a choice from among them, and then implementing that choice. Decisions also involve a time element: The right decision is the best solution at the time it must be made (Fig. 12.2).

Figure 12.2 illustrates a typical issue management procedure.

12.2.5 Change Control Management: Broad Approach

A change is defined as a requested modification at any stage of the project, which requires the deliverable under development to deviate from the current baseline.

The business requirement mapping document will serve as the baseline for user requirements in a project. Any deviation from the proposed system is defined as a

12.2 Project Management Life Cycle

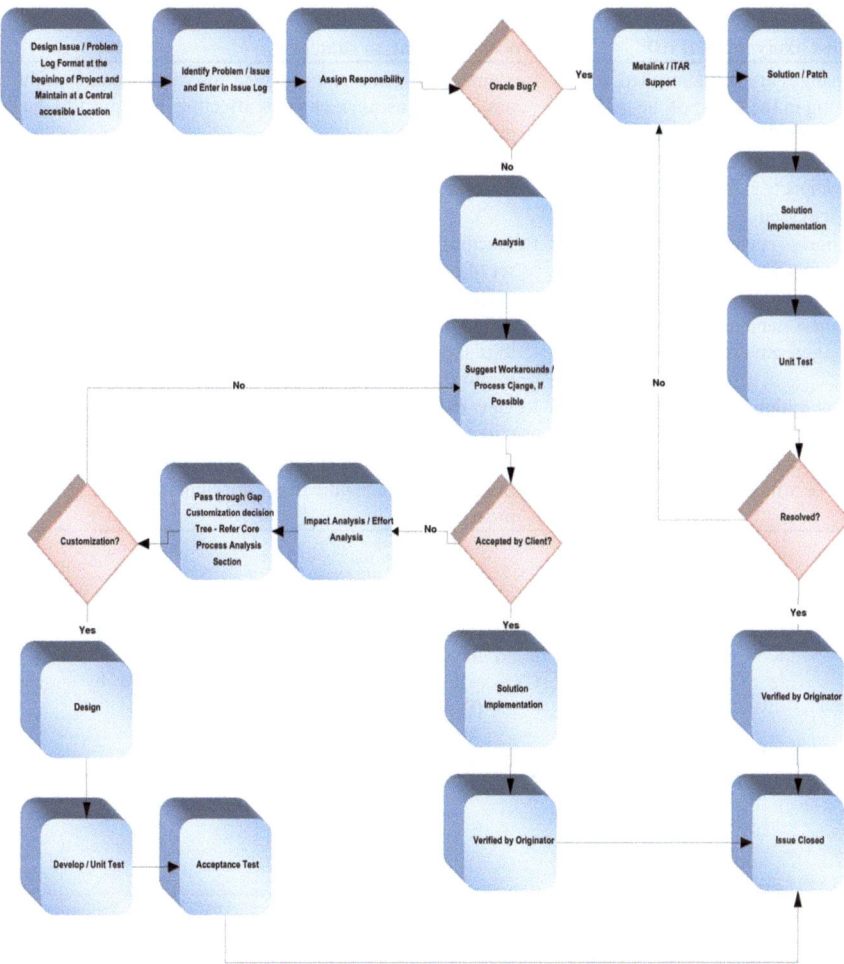

Fig. 12.2 Typical issue management procedure

change and will be subjected to the change management procedure outlined later in this section.

The change management procedure will cover:

- Identification and documentation of the need for change and change details (request for change).
- Analysis and evaluation of change request for technical impact and impact on effort, schedule, as well as other planning aspects.
- Approval or disapproval of change request.
- Implementation of change.

Table 12.3 Most likely area of risk and probable mitigation strategies

Most likely areas of risk	Probable mitigation strategies
Changes are not controlled, resulting in the scope of the project gradually increasing	Implement and enforce the change control procedure on the project, and escalate changes affecting the project to the project sponsor
Issue, change, and problem impacts are not adequately assessed	Ensure that impact assessments are thoroughly prepared and reviewed
Progress reviews are irregularly or superficially conducted	Be uncompromising in insisting on a regular and complete set of project reviews, both within the project and to project stakeholders
Project management does not aggressively resolve issues	Hold regular issue review meetings and review issue backlogs at project progress reviews. **Use a** *top ten issue* **list to maintain visibility of the most critical issues**
Risk mitigation strategies are not monitored for effectiveness	Review risks regularly at project progress reviews with the client and consulting management
Deliverables are not baselined or controlled	Install configuration management repository tools like Visual SourceSafe, etc., and enforce configuration management procedures

Table 12.4 Log table

Issue #	OA module	Issue description	Status	Priority	Date raised (C)	Aging days =IF (B=0, A−C, B−C)	Raised by	Assigned to	Resolution plan	Actual resolution	Closed date (B)	Verified by	Verify date
–	–	–	–	–	–	–	–	–	–	–	–	–	–
–	–	–	–	–	–	–	–	–	–	–	–	–	–
–	–	–	–	–	–	–	–	–	–	–	–	–	–
–	–	–	–	–	–	–	–	–	–	–	–	–	–

The table describes a typical format of an issue log
A today

The change request may be raised by users or by any of the project team members.

All approved documents from the client will be considered final on receipt by the consultant. Any changes required for a final document must go through a formal change management procedure. They should also be logged as *change management requests (CMRs)*. Either client or consultant may raise a CMR. The impact in terms of time, cost, and effort will be evaluated by the consultant and will be submitted to the client and get authorization for the change implementation. The CMR should be validated by a strong business case as to why such a change is required to be implemented.

Typically, a change control request form will contain the following:

- Details of change
- CMR number of the change
- Date originated

12.2 Project Management Life Cycle

- Who originated (client/consultant)
- Description of change
- Criticality/priority
- Assessment of change
- Description of change
- Business case justifying the change
- Technical impact
- Estimated effort and cost
- Impact on current schedule
- Decision
- Authorization
- Schedule for carrying out CMR

A change control register will be maintained to track the status of CMRs and the progress of their implementation.

The change control register will contain the following:

- CMR number
- Priority/criticality
- Date originated
- Who originated (client/consultant)
- Date issued
- Date authorized
- Scheduled start date
- Scheduled completion date
- Actual start date
- Actual completion date
- Completed by whom
- Verified by whom
- Status of CMR (completed/partially complete, and not started)

The status of the CMR will appear in the monthly status reports of the project leader (consultant) to the client.

12.2.6 *Status Monitoring and Control: Broad Approach*

The single most critical task during project control is status monitoring and reporting. If properly employed, the regular progress reporting and review cycles can provide a solid forum from which the project manager can stay in control of project work. This activity will receive and create a wealth of qualitative and quantitative information in these reviews to help the project team find problems and fix them before they become uncontrollable.

Team Progress Reviews should be held on a weekly basis to assess the progress of each team and to plan for the following week or weeks. They also include a discussion of any issues and problems. The project manager should chair the meeting.

Project Progress Reviews are normally held at monthly intervals, covering each financial reporting month. The project manager gives a report to the client and consulting management, summarizing progress, problems, risks, issues, and any proposed changes.

The Project Steering Committee meets at an interval usually determined by the project sponsor to review the progress reports and discuss business issues relevant to the project. The project manager, the client project manager, and usually the product vendor representative represent the project at these meetings. The consulting business manager should be present at this review. The committee will discuss:

- Major issues needing immediate action
- Change requests needing immediate action
- Changes to the schedules, if any
- Problems encountered and corrective actions taken
- Any decisions on technical matters

The consultant project leader will also provide a monthly status report to the client. This report will highlight the following:

- Activities completed against planned
- Milestones achieved
- Changes to the schedules, if any
- Problems encountered and corrective actions taken
- Outstanding problems and issues
- Plan for implementation of approved change requests and status of
- Current change requests
- Plan for the next period

12.2.7 Problem Reporting and Escalation

The consultant project leader will resolve any problem that needs resolution with the client project manager.

The project leader in consultation with the client will decide on the action to be taken on the issue. If it needs the attention of higher management, it will be reported to the steering committee.

The steering committee shall decide on the action to be taken on the issue. If it needs the attention of higher management, the problem will be accordingly reported to the senior management.

12.2 Project Management Life Cycle

Table 12.5 Processes for project completion

Process Ref No.	Project management Task
1	Secure client acceptance of all project deliverables
2	Release physical resources from the project
3	Audit key deliverables
4	Conclude configuration management and transfer the repository to the client
5	Conduct project closure and client acceptance letter

Table 12.6 Deliverables and responsibility

Sl No.	Name of deliverable	Responsibility	
		Major	Assisted
1	Unified Project Plan		
2	Problem/issue log format	Consulting organization	Client
3	Change management request format	Consulting organization	Client
4	Project progress report format	Consulting organization	Client
5	Internal quality review format SQAP	Consulting organization	Client

SQAP software quality assurance plan

12.2.8 Project Completion

The goal of project completion is to secure client acceptance of the project and to close the project in an orderly fashion. A well-managed project completion demonstrates that you have met your client's needs and allows you to negotiate a successful conclusion to the project. A satisfied client will act as a reference and a potential source of future engagements (Table 12.5).

The objectives of project completion are to:

- Gain client acceptance of all project deliverables
- Close out the contractual agreement, if any, with the client
- Hand over project deliverables and environments to the production support team (as appropriate)
- Release staff and physical resources
- Document and archive project results in the consulting practice

12.2.9 Manage Acceptance Expectations Carefully

When conducting project completion, remember to continue to manage changes, issues, and problems throughout acceptance. Last-minute issues and problems can be quite common, as client stakeholders realize that they have only a short time remaining to influence project results. The project sponsor and client project manager should take a leadership role at this point in the project to control the introduction of new issues from the client.

Table 12.7 Major activities, checkpoints, and weightage

Major activities	Checkpoints	Weightage
Project planning	Refer "Project Initiation" section	
Project control	Are the variances against project plan being reviewed and corrective action identified?	
	Are corrective actions being taken and issues closed as per issue log?	
	Are the configurable items controlled and under strict configuration management?	
	Are all changes to scope and enhancement requests being documented and evaluated through CMR?	
	Is the resource staffing in corresponding phases correct and the project team not being overburdened to meet target?	
Project completion	Have all deliverables been signed off?	
	Are all critical issues as per the issue log closed?	
	Have all the change requests as per CMR been closed?	
	Is the client unhappy about any unresolved issues?	

12.3 Deliverables

12.4 Decision Matrix/Checklist

12.5 Critical Success Factors

The critical success factors of managing a project smoothly and successfully are as follows:

- Scope, objectives, and approach are agreed upon and understood by all parties.
- Project culture and climate are established conducive to a win–win philosophy.
- Risks are identified and containment measures are put in place.
- The client understands the obligation to provide resources to support the project work plan.
- Work progress is reviewed regularly and aggressive action is taken to correct variances to the work plan.
- Project deliverables are protected from unauthorized change and baselines are fully compliant with project requirements.
- Good client relationship and managed expectations.
- Outstanding issues and problems resolved prior to their acceptance.

Chapter 13
ERP in B School

13.1 Introduction

Management education has been traditionally conducted by business schools that focus on imparting a broad range of managerial knowledge and abilities. Initially, developing a robust curriculum that would set challenging atmosphere and create near-perfect simulated conditions for the aspiring managers was considered as the differentiation point among the B schools (Wexley and Baldwin 1986). However, with passage of time, curricula adopted by different B schools are near similar; this has been possible because of the advent of open-source Internet where many materials have been shared with the intent to make education all pervasive. In the present knowledge economy, however, the business dynamics are changing very fast, and before a curriculum can be standardized, the paradigm would have shifted to another level.

To meet these changes, not only should the curriculum be agile and dynamic but other associated processes related to B school education should also be able to seamlessly integrate and produce the desired output to budding managers. Reshaping the present management education to the emerging needs of the business management will require not only addressing course curriculum but also integrating several academic processes, such as designing new courses, faculty recruitment and skill development, improving infrastructure and admitting the right mix of student profile.

13.2 Literature Review

Efficient and capable managers are crucial to the success of an organization. As we are facing rapid changes in the business scenarios, these managers become the link for successful design and execution of a strategy (Tannenbaum 2002). To remain agile and retain the competitive advantage, these managers need to achieve results against a background of intensified competition, incessant change, demanding customers, managing eco-partners (such as suppliers and distributors) across geographical boundaries and still be able to reduce attrition. If B schools have to make their students ready for such a scenario, they need to achieve greater integration of curriculum, pedagogical efforts, faculty development and innovation (McCuddy and Pirie 1998).

In addition to these requirements, the B schools also need to train and keep their students updated on information technology as information systems have taken a role of strategic position from being just an enabler (Hammer and Champy 1993). This means that the students from B schools need to be trained on the usage of technology (Pfeffer 1977) along with traditional concepts. This will transform these budding managers as strong human and social capital in the organizations that they would be employed in. Leonard-Barton (1995) argued that a holistic knowledge can be imparted to students if there is an effective integration of traditional concepts, technology and all other academic processes. Thus, it is evident that there is a need for managers with holistic skills and knowledge who can manage new complex situations, partnering with eco-partners and customizing in global markets (Conger et al. 1999). Hence, according to McDonald and Mansour-Cole (2000), there is a gap between the supply and demand for the capable managers, and B schools need to provide the required inputs to make their students business leaders in the knowledge economy era.

Management education needs to address the above-mentioned challenges by exposing their students to these challenges by simulating these situations in their course curriculum (Hallinger and Snidvongs 2008). There is a growing awareness that the management students need to be developed in an appropriate and desired way so that survival and future success of organizations are ensured (Buckley and Kemp 1987). However, it is not easy for B schools to rise and address all these challenges without putting enough effort. Spender (1995) feels that B schools are not up to the mark and the similar opinions have been echoed by Hambrick and Jorgensen (Hambrick 1994; Jorgensen 1992) in their research works. Management teachers and educationists are increasingly facing the challenge to transform their students to the needs of the corporate, while braving increased competition for international rankings to meet the needs of funding agents for different projects (Huff and Huff 2001). According to Forman (1998), the B schools, apart from meeting the needs of the corporate also have to meet the needs of different stakeholders (employees, advisory boards, funding agencies, alumni groups and students).

The review of literature suggested that there is a need for the B schools to change its approach. It also points out several goals that need to be met to cope with the changing business environment. However, traditional functional curriculum does not address these issues (McCuddy et al. 1998) and needs to move towards coherent and developing students who can understand the role that each function plays in furthering an organization towards meeting its business goals. However, the existing literature does not talk about a specific approach or framework for transforming their students into coherent and holistic managers.

13.3 Objective

In view of the gap in the literature, the present study has the objective to develop an integrated approach which can help B schools to transform their students into effective managers who can face the vagaries of changing business dynamics and complex situations and meet the needs of all the stakeholders.

13.4 Methodology

The methodology for collecting data has been through secondary research in which we have studied the approach taken by all top business schools in India. The top business schools were taken from the published ranking by Credit Rating Information Services of India Limited (CRISIL; http://www.crisil.com/ratings/bschool-school-grading-list.html). A*** ratings were provided to top B schools. We have studied the approach taken by the Xavier Institute of Management, Bhubaneswar (XIMB; www.ximb.ac.in) as it was easier for the authors to collect data from the institute and this institute responded to our requests for providing information for the research.

13.5 Case Study XIMB

XIMB is one of the premier business schools in India established in 1987. It traces its origins to a unique partnership between the Government of Orissa and the Jesuit Society which is widely recognized for its contribution to higher education in India.

The institute is situated in a rapidly evolving software hub in Eastern India. The institute functions from a technology-intensive campus in the city of Bhubaneswar. It has developed strong ties with Confederation of Indian Industries (CII). It has been South Asian Quality Assurance System (SAQS) accredited by the Association of Management Development Institutions in South Asia.

XIMB has collaborated with many B schools in North America, Europe and Africa. Through industry-relevant projects, the institute facilitates high-impact business research. The faculty is involved in a lot of research publications. It publishes academic journals, such as *Vilakshan* and *The Journal of Research Practice*.

It offers a lot of academic programmes to produce people who can impact the world of business with their skills, knowledge, and pursuit of excellence, but with a human touch:

- Post-Graduate Diploma in Management (PGDM-BM) is the flagship course
- Post-Graduate Diploma in Rural Management (PGDM-RM)
- Post-Graduate Diploma in Business Management (for working executives) (PGDM-Part time)

- Fellow Programme in Management (FPM)
- Post-Graduate Certificate Programme in Business Management (PGCBM) through video conferencing
- Post-Graduate Programme in Management and Insurance (PGPMI)

The institute regularly invites renowned academicians, diplomats, resource persons, industrialists, as well as top-level executives who regularly address students, teach courses, collaborate in research projects, as well as participate in management development programmes (MPDs), research seminars and other academic pursuits.

13.5.1 A Step Forward by XIMB

In this fast-evolving world, business is changing by the minute. Business requirements are becoming demanding. Managing business is increasingly becoming complicated and cumbersome. It is a specialized job no doubt, and people need to be trained for that. There is a huge demand for management education in the country and away. Yet quality supply is not available abundantly. The demand far outstrips the supply. While there is a lot of young talent out there willing to take up formal management education, many are unable to do so due to lack of time, workplace limitations and location inconveniences.

Organizations consist of two kinds of white-collar jobs: those with technical skills and those with managerial skills. The technical person is concerned about doing an assigned task in a set way. He/she does not always look at the overall process whether it is improving a certain parameter. On the other hand, for the person responsible for managing the job, if he/she knows the proper technical requirements for setting certain tasks for the technical guy, then he/she can set things straight. But if he/she does not know, then things can go awry. Quite often, there is a disconnect between what is required and what is actually delivered. This leads to what Wickham Skinner calls the *millstone effect*. The technical person takes more important decisions than is expected of him/her. So it is essential to equip people in the technical field with managerial knowledge.

To meet this ever-growing demand for managerial education, XIMB started the PGCBM through video conferencing in 2005. It was among the first premier management institutes in the country to understand the demand for distance learning management education. The programme is designed to provide working professionals and managers with a broad overview of the concepts, various tools and techniques needed to meet today's business challenges.

Following the early footsteps taken by XIMB, other institutes in the country have also started similar courses. Some of the Indian Institutes of Management (IIMs) have also started similar courses. The Indian Institute of Management Bangalore (IIMB) offers the Advanced General Management Programme (AGMP) and Executive General Management Programme (EGMP) through satellite medium.

Benefits

The PGCBM course offers a lot of opportunities. First and foremost, it helps bridge the huge demand for management positions. Earlier, there were people who could not be targeted even though there was a demand for it because of certain constraints. But this course cuts across barriers of space, time and location. Students get a boost and their outlook towards management changes. The professors learn to handle and manage forum discussion since that is an important arena for students to discuss among themselves. This helps the professors to moderate in the forum discussion of full-time students. Moderating in the class and moderating in the discussion forum are two entirely different experiences and require different approaches to be tackled with.

Since many of these people are involved in technical tasks, companies can fill up many positions which require people with managerial skills and having a technical bent of mind.

13.6 Analysis

The case shows that there are four factors that are important for a holistic management education: feedback from stakeholders (including recruiters), structure of the course based on the needs, ease of access (using technology) and flexibility to take care of changes required. There are roles that are played by administrators, staff, infrastructure department (particularly computer resource centre), corporate recruiters (who define their needs as well as provide feedback on a continuous basis), besides faculty who can make a course an acceptable one. The approach has been a hybrid one where part of the course has been delivered using traditional pedagogy and the rest using online, virtual mode. The results identified by the approach have been found to be beneficial to all stakeholders. The approach is diagrammatically represented in Fig. 13.1.

13.7 Conclusion

As it came clearly in the XIMB case study that integration creates a lot of value. The online model used by XIMB was the first step towards bridging a gap between the industry and education institute. The profile of students is quite rich. They came from various corporate backgrounds. XIMB has created a platform in which they can share their experience and knowledge not only among themselves but also with faculty. It results in relevant, up-to-date case studies which can be discussed among regular XIMB students. These students are getting exposure to real issues in organizations and solving these issues in a safe environment. While designing the curriculum, several other processes need to be integrated with the delivery of lectures, such

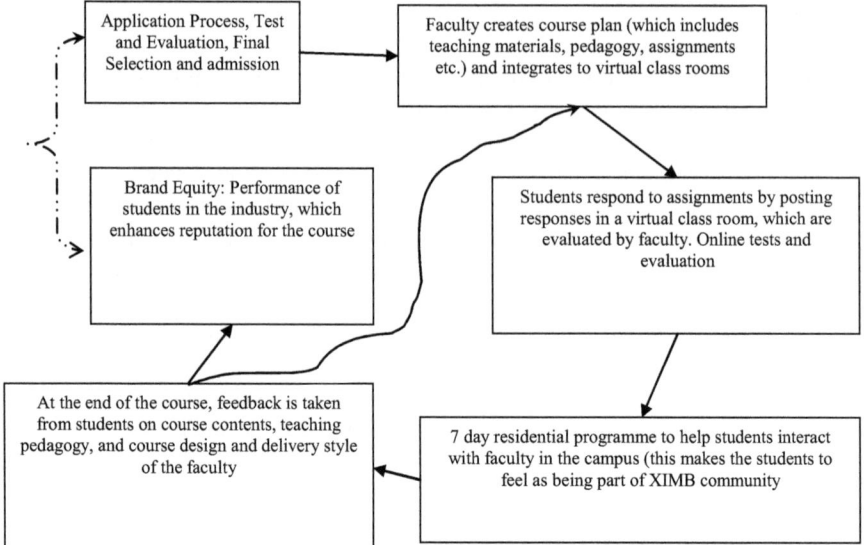

Fig. 13.1 Framework for using information system in a holistic manner

as scheduling, resourcing, faculty selection, online portal availability, etc. These integrations help to make it a seamless delivery of the management education and still be able to improve, if required. It also helps faculty to conduct research in the areas which popped up during discussion with online education-experienced students. Finally, it also brings new consulting assignments to faculty. Therefore, these organizations also get benefitted by a professor's expertise. In other words, XIMB has set up a knowledge-sharing platform among faculty, students, consultants and organizations.

The next big question is how we can use innovation to further improve graduate management education and integrate it with society. What technology can the faculty use to obtain the greatest level of dialogue among all stakeholders? However, this case clearly demonstrates the untapped value which arises from integration.

Further Reading

Adria M, Teresa R (2004) Technology, preprocessing, and resistance—a comparative case study of intensive classroom teaching. J Education Bus 80(1):53–60

Ahuja G (2000) Collaboration networks, structural holes, and innovation: a longitudinal study. Adm Sci Q 45(3):425–455

Anderson K (1999) Internet-based model of distance education. Hum Resour Dev Int 2(3):259–272

Behrman JN, Levin R (1984) Are business schools doing their job? Harv Bus Rev 1984(January/February):140–147

Bettis RA, Hitt MA (1995) The new competitive landscape. Strateg Manag J 16:7–19

Buckley J, Kemp N (1987) The strategic role of management development. Manag Education Dev 18(3):158–159
Conger JA, Spreitzer GM, Lawler EE (1999) Introduction: the challenges of effective change leadership. Jossey-Bass, San Francisco
CRISIL http://www.crisil.com/ratings/bschool-school-grading-list.html
Cunha MP et al (2004) Looking for complication: four approaches to management education. J Manag Education 28(1):89
Dacko SG (2001) Narrowing skill development gaps in marketing and MBA programs: the role of innovative technologies for distance learning. J Mark Education 23(3):228–239
Davis HJ (1998) A review of open and distance learning within management development. J Manag 15(4):20–34
Forman S (1998) Undergraduate education reform. Paper presented at the AACSB undergraduate Program Seminar Charlottesville, VA
Gibson JW, Heerera JM (1999) How to go from classroom based to online delivery in eighteen months or less: a case study in online program development. T H E J 28(8):57–60
Goodrick E (2002) From management as a vocation to management as a scientific activity: an institutional account of a paradigm shift. J Manag 28(5):651–655
Hall OP Jr (2006) Enhancing management education using hybrid learning nets: a perspective from working adults. J Bus Manag 12(1):45–58
Hallinger P, Snidvongs K (2008) Education leaders: is there anything to learn from business management? Educational Manag Adm Leadersh 36(1):11–12
Hambrick DC (1994) 1993 Presidential address: what if the academy actually mattered? Acad Manag Rev 19(1):11–16
Hamilton D et al (2000) A decision model for integration across the business curriculum in the 21st century. J Manag Education 24(1):103
Hammer M, Champy J (1993) Reengineering the corporation. Harper Collins, New York
Hay A, Hodgkinson M, Peltier JW, Drago WA (2004) Interaction and virtual learning. Strategic Change 13(4):193–204
Huff AS, Huff JO (2001) Re-focusing the business school agenda. Br J Manag 12:S49
Jorgensen B (1992) Industry to B. schools: smarten up on TQM or else. Electron Bus 18:85–90
Kearns KP (1998) Institutional accountability in higher education: a strategic approach. Pub Product Manag Rev 22(2):140–156
Latham Gary et al (2004) Fostering integrative thinking: adapting the executive education model to the MBA program. J Manag Education 28(1):3–9
Leonard-Barton D (1995) Wellsprings of knowledge: building and sustaining the source of innovation. Harvard Business School, Boston
Linder JC, Smith HJ (1992) The complex case of management education. Harv Bus Rev 70(5):16–33
McCuddy MK, Pirie WL (1998) Contributing to a skilled workforce: Val-paraiso University's approach. Paper presented at the AACSB undergraduate Program seminar, Charlottesville, VA
McDonald KS, Mansour-Cole D (2000) Change requires intensive care: an experiential exercise for learners in university and corporate settings. J Manag Education 24(1):128
Pfeffer J (1977) Effects of an MBA and socioeconomic origins on business school graduates' salaries. J Appl Psychol 62:698–705
Porter LW, McKibbon LE (1988) Management education and development: drift or thrust into the 21st century? Mc-Graw Hill, New York
Quinn RE, Snyder NT (1999) Advanced change theory: culture change at Whirlpool corporation. In: Conger J, Spreitzer GM, Lawler EE (eds) The leader's change handbook. Jossey-Bass, San Francisco, pp 162–194
Raelin JA (2000) Work based learning: the new frontier of management development. Prentice Hall, Upper Saddle River
Shea T, Motiwalla L, Lewis D (2001) Internet-based distance education—the administrator's perspective. J Education Bus 77(2):112–117
Spender JC (1995) Underlying antinomies: an historical analysis of US graduate business education and its problems. http://www.iris.nyit.edu/~spender. Accessed in April 2013

Tannenbaum SI (2002) A strategic view of training and learning. In: Kraignr K (ed) Creating, implementing, and managing effective training and development: state-of-the-art lessons for practice. Jossey-Bass, San Francisco, pp 10–52

Volery T, Lord D (2000) Critical success factors in online education. Int J Education Manag 14(5):216–223

Vrasidas C, Zembylas M (2003) The nature of technology-mediated interaction in globalized distance education. International. J Train Dev 7(4):271–286

Wang AY, Newlin MH (2001) Online lectures: benefits for the virtual classroom. T H E J 29(1):17–21

Wexley KN, Baldwin TT (1986) Management development. J Manag 12(2):278

XIMB http://www.ximb.ac.in

Index

A
Activity flow diagram, 40
Analyze dependency, 85, 87, 89
Application setup, 71, 73, 76, 77, 103
  definition of, 75

B
Backup strategy, 74, 82
Business
  flow diagram, 43, 91, 73, 92, 101, 102, 73
  requirement definition, 46
  requirement mapping, 75
    documentation, 154
  system testing, 78, 79, 101, 103, 106
    objectives of, 78

C
Change control management, 154
Change management, 37, 41, 155, 156
Completing tasks
  assumptions, 43
  client policies, 44
  communicate with honesty and conviction, 43
  consultant proposal, 44
  consulting policies, 44
  contract review, 44
  contractual agreement documents, 44
  define the project, 42
  environment requirement, 44, 74
  exclusions, 43
  field a project team, 44
  field a winning team, 42
  firm-up project scope, 43
  install the software, 44, 95, 97
  know your client, 42
  manage the risks, 42
  plan for completion, 43
  produce formal documentation, 42
  resource costs, 44
  unified project plan, 41, 44, 123, 160
Conference room pilot, 37, 40, 104
  definition of, 71
  objective of, 71
Configuration management, 41, 126, 154
Control and reporting, 41, 151
Controlling tasks, 42
Core process analysis, 37, 40, 87
Critical
  setup, 76
  success factor, 37, 40, 47, 82, 89, 99, 106, 127, 160
CRP instance, 71, 73, 75, 77, 82
Customization, 85
  design/build, 40

D
Data migration, 11, 91, 97
Decision matrix/checklist, 82, 127
  a prerequisite to proceed to the next phase, 40
  to gauge successful completion of the phase and a prerequisite to proceed to the next phase, 40
Design, 1, 18, 37, 45, 74, 88, 89, 92, 94
Detailed activities containing the tasks corresponding to that activity, 40
Development, 17–19, 24, 26, 44, 74, 75, 88, 92, 94, 95, 105
  framework, 27
  skill, 161
  sustainable, 7

E

ERP, 9
  application standard functionalities, 86, 93
  design processes, 2
  evolution of, 2
  implementation processes, 8
  modules of, 8
  overview of, 7
  phases, 2

G

Gap
  analysis, 71, 87, 89, 90
  resolutions, 87
Go live, 20, 22, 37, 40, 98

I

Implementation schedule, 40, 44, 73, 92, 93, 124, 73
Instance strategy, 71, 73, 82
  objective of, 74
Interface
  analysis, 93, 98
  building team, 93
  program, 93
Interface/Conversion design/Build, 40

P

Parallel process, 82
Phase deliverables, 40
Planning tasks, 42, 153
Prerequisites, 39, 40, 44, 46
Production Go-Live, 37, 41
Project completion, 43, 153, 159
  objectives of, 159
  processes of, 159
Project initiation, 37, 40–43, 45, 46, 153
  objective of, 41
Project management, 8, 42
  approaches to, 41
  categories of, 151
  goals of, 41
  objectives of, 151
  tools, 8

Q

Quality management, 8, 41, 151

R

Reporting, 41, 43, 45, 104, 157, 158
Resource management, 8, 41, 45
Risk management, 44, 154

S

Source data files, 97
Status monitoring, 157
Steering Committee, 125, 158
System
  administration, 77
  integration testing, 41, 101, 104, 105

T

Testing, 11, 24, 37, 71, 74, 78, 79, 82, 96, 101, 102–104, 123
  of products, 24
  process, 78, 101
  requirements and strategy, 79
  unit test, 80
To Be process, 10, 86, 87, 93
Training, 11, 14, 25, 37
  cost of, 20
  use of skilled users for, 124
Transition, 16, 18, 39

U

UAT
  conduct, 126, 127
  environment, 124, 125
  execution plan, 125
  strategy, 124, 125
Unit Testing and CRP, 78, 79
User acceptance support, 41
User acceptance testing, 37, 96, 97, 125, 126
  processes for, 125
User requirements
  baseline for, 154

W

Work management, 41, 151

MIX
Papier aus verantwortungsvollen Quellen
Paper from responsible sources
FSC® C105338

If you have any concerns about our products,
you can contact us on
ProductSafety@springernature.com

In case Publisher is established outside the EU,
the EU authorized representative is:
**Springer Nature Customer Service Center GmbH
Europaplatz 3, 69115 Heidelberg, Germany**

Printed by Libri Plureos GmbH
in Hamburg, Germany